DIGITAL
ELECTRONIC
CIRCUITS
The Comprehensive View

DIGITAL ELECTRONIC CIRCUITS

The Comprehensive View

Alexander Axelevitch

Holon Institute of Technology, Israel

World Scientific

NEW JERSEY · LONDON · SINGAPORE · BEIJING · SHANGHAI · HONG KONG · TAIPEI · CHENNAI · TOKYO

Published by

World Scientific Publishing Co. Pte. Ltd.

5 Toh Tuck Link, Singapore 596224

USA office: 27 Warren Street, Suite 401-402, Hackensack, NJ 07601

UK office: 57 Shelton Street, Covent Garden, London WC2H 9HE

Library of Congress Cataloging-in-Publication Data
Names: Axelevitch, Alexander, author.
Title: Digital electronic circuits : the comprehensive view / Alexander Axelevitch.
Description: New Jersey : World Scientific, [2018]
Identifiers: LCCN 2018027359 | ISBN 9789813270725 (hardcover)
Subjects: LCSH: Digital electronics. | Digital integrated circuits. | Logic circuits.
Classification: LCC TK7868.D5 .A95 2018 | DDC 621.381/32--dc23
LC record available at https://lccn.loc.gov/2018027359

British Library Cataloguing-in-Publication Data
A catalogue record for this book is available from the British Library.

For any available supplementary material, please visit
https://www.worldscientific.com/worldscibooks/10.1142/10998#t=suppl

Desk Editor: Tay Yu Shan

Typeset by Stallion Press
Email: enquiries@stallionpress.com

Preface

This book deals with key aspects of designing and building digital electronic circuits on the basis of the different families of elementary electronic devices. The implementation of various simple and complex logic circuits is considered in detail, with special attention paid to the design of digital systems based on complementary metal-oxide-semiconductor (CMOS) and pass-transition logic (PTL) technologies acceptable for use in planar microelectronics technology. It is mainly intended for students in electronics and microelectronics, with exercises and solutions provided.

Alexander Axelevitch

Introduction

The main purpose of this book is to help present and future students understand the principles of digital circuits design, and their place and application in microelectronics. The second goal, which may be more important for students, is to help them overcome the exam for the course "Digital Circuits" which I taught at the Holon Institute of Technology (HIT) for more than 20 years, two semesters per year, including the summer semester. Considering that after each semester I prepared three variations of the exams, I was compelled to always come up with new schemes, questions and problems. I present most of these examination problems together with the solutions in the book as examples. So, I hope that this book will be useful to students, as well as practical engineers engaged in the design and realization of microelectronic circuits and devices.

The course "Digital Circuits" is part of a system consisting of several consecutive courses: "Digital Logic Design," "Digital Electronic Circuits," "Digital Systems," "VLSI Design," and "Programming with the VHDL." Evidently, this sequence can provide an understanding of only one side of microelectronics for students. The second side is the technology. However, the part devoted to the design is interesting and important. The course "Digital Electronic Circuits" allows you to give meaning to those "black boxes" that comprise logic, and fill them with diodes, transistors, and other elements. A complete study of these systems enables you to understand and create novel digital electronic devices for different applications: the storage, processing and transmission of the information presented in digital form, integrated sensors and control devices, computers and microprocessors.

Every novel field of science is based on its specific language first of all. So, we need to begin to study the design of digital electronic circuits by studying its novel language — which entails learning new descriptions and definitions. This language consists of defined and enough formal constructions based on the Boolean logic. These structures are partially based on the language of mathematical logic.

The book is organized as follows: the first part of the book deals with the general definitions and basic parameters characterizing the logic of digital circuits, and the transfer function definition. The second part is dedicated to logic families based on the bipolar semiconductor devices. The third part is devoted to logic families based on the unipolar semiconductor devices and advanced technologies applicable in the field of digital circuits. The fourth chapter of the book deals with the analysis and synthesis of digital logic circuits. The basic principles of the synthesis of combinational logic circuits using CMOS logic, dynamical logic, and pass-transistor logic are considered in this part. Part 5 of the book is devoted to sequential logic: flip-flops, registers and counters, multivibrators. The last chapter is dedicated to the semiconductor memory architecture and related combinational digital circuits. Each part of the course will be accompanied by exercises supplied with full solutions and explanations that aim to make the learning more fruitful.

This book can serve as basic study material for the course "Digital Electronic Circuits." It is intended for students studying the specialties related to electronics and microelectronics, as well as for practical engineers working in the field. The author wishes all future readers and users of this book a pleasant read.

Contents

Chapter 1

Basic Definitions and Logic Families

1.1. Basic definitions

Every physical process or event, every situation and condition, may be characterized numerically with the help of certain signals or quantities. These quantities may be observed, measured, manipulated, and stored (recorded). A quantity may be represented in analog and digital forms. Usually, all the signals existing in nature are analog or continuous values. Temperature, for example, changes continuously. The water level in a chemical reactor changes continuously also. If the signal varies continuously, we can say that its magnitude can take all possible values in the domain of the definition of the signal.

Analog signal processing can be performed by using different instruments. It can be performed by both analog and digital devices. However, to process an analog signal with a digital device, using a digital computer for example, we should convert the analog signal to digital form. To convert the signal we must measure it. We provide our measurements using existing measuring devices which have a defined accuracy. Moreover, we provide our measurements at defined intervals of time. An operation providing the analog–digital signal conversion is called **sampling**. Figure 1.1 illustrates this type of conversion. As shown in Fig. 1.1, an analog signal varies in time. We measure it at the points t1, t2... using a specific measurement device with suitable precision. All measured values F1, F2... are

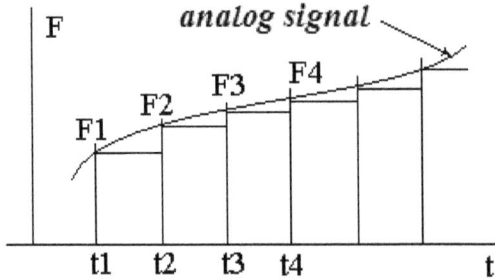

Fig. 1.1.   A sampling operation to convert the analog signal to a digital one.

proportional to some minimal value called "unit," which defines a range of approximation or a resolution of the device. The obtained quantity of measured units presents a numerical form of the measured signal which may be presented in one of the numerical digital scales, for example in the decimal or binary form.

This form is very convenient to process in digital computers. However, each numerical form always contains a systematic error. For example, we can always insert additional numbers between two neighbors. So, each signal may be presented in the analog as well as digital form. Now we can define the concepts of analog and digital signals:

- An **analog value** can have any level within certain operating limits, as long as it is proportional to the signal.
- A **digital value** can only have a number of fixed values within certain tolerance limits. Here, the quantity is represented by symbols called **digits**.

Figure 1.2 illustrates the principle of analog and digital signal measurement. On the left side of the picture, the analog type of measurement appears. The level of water in the tank changes continuously and we can draw a graphical representation of the water level as a function of processing time. On the right side, we can see the digital or discrete measurement system. The threshold of the water level is fixed. If a float comes into contact with the fixed contact representing the threshold, a relay will switch from the 0 state to the

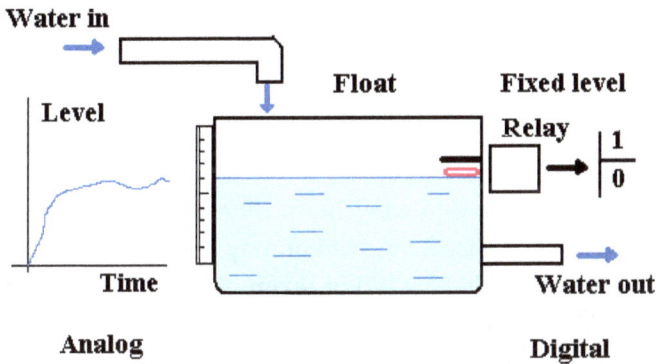

Fig. 1.2. A simple measurement system.

Fig. 1.3. Various measurement systems.

1 state and will signal the level change. In this digital binary system, we can measure two levels: lower and higher.

Figure 1.3 shows two additional examples of the measurement of different quantities.

In the picture on the left, a thermocouple (two different metals joined together) measures the temperature and transforms this natural parameter into voltage. All changes are continuous in the system; therefore, such a measuring system implements an analog device. In the picture on the right, a lamp shows the state of the switch "S." If a supply is connected to the lamp by the switch, the lamp lights up and we can see that "S" is in the connected state. The switch "S" only has two different states. In other words, we can describe the action of this switch with two digits, 0 and 1. A **digital system** is one that processes a finite set of data in a digital (discrete) form. The signals may be represented in **binary** form or in a form that only has two **operating states**. These operating states, called **logic states**, may be of high (1) or low (0) **logic levels**. An electrical or electronic circuit implementing the possibility of obtaining two different logic levels opposite to the input level is called an **inverter**. An inverter is a minimal **logic gate**. So, a logic gate is a standard electronic scheme implementing some logic operation, logic sum or logic multiplication, for example.

The digital system that may be in different logic states is called the **logic system**. Every logic system may be represented by a finite number of discrete elements. For example, the Hebrew language consists of 22 letters, English has 26 letters, and Russian has 33 letters. However, each of them can put into words all the nuances and inflections of human sensations and thoughts. Thus, each logic system may be implemented in numerous different ways. The next example is as follows: ten decimal digits permit the creation of mathematics... Two or three digits permit the creation of mathematics also. So, one can say that the discrete appearance of the natural analog function is ambiguous; there are a multitude of right solutions for digital logic systems. We will see further that various logic systems or electronic schemes may implement the same logic function. The art of logic design consists of the knowledge of all the methods and techniques and the ability to apply them to create the optimal solution. What criterion of the optimization, it will be clear from specific conditions. That may be a cost of devices, or processing rate, or some other reasons.

A logic system consists of logic devices. A connection of two simple logic devices is presented in Fig. 1.4.

A logic system may be loaded by a few logic gates. Logic gates may be created using various technologies: diodes, bipolar transistors, MOS transistors. They may be electronic or pneumatic devices. Usually, logic gates from the same family of devices are used for the creation of complex logic systems. Figure 1.5 illustrates a connection of the same logic gates in one more complex system. A **fan-out** is the number of logic gates connected in parallel with the output of the logic gate driving these gates, on condition that all these devices belong to the same logic family. The maximum fan-out is equal to the number of logic gates which can load the driving gate before its performance is impaired. A **fan-in** is a number of logic gates which

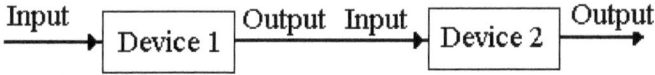

Fig. 1.4.   A connection between two logic devices.

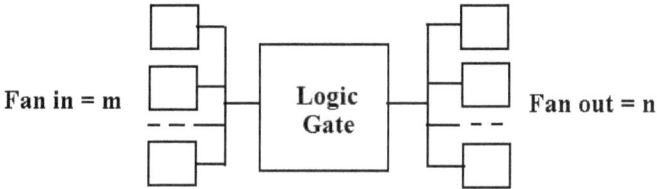

Fig. 1.5.   The definition of input and output loading in the logical systems.

are connected in parallel with the input of a logic gate from the same logic family. A maximum fan-in is defined as the number of logic inputs which can be accommodated by a logic gate from the same logic family. In some logic families the operating speed falls as the number of inputs is increased. So, the values of fan-in and fan-out are significant parameters of a logic system and the designer must take into account all limitations and conditions of the designed devices.

A very good example of the fan-out value is the number of memory cells that may be driven by one controller. These parameters of logical systems, fan-in and fan-out, are required at a stage of technical design. However, the logic designer has to consider the type of technology which he will use while creating the system.

All logic systems may be divided into two basic groups: **combinational logic systems** and **sequential logic systems**. Figure 1.6 presents the basic scheme of the combinational logic circuit. The main property of combinational logic circuits is the dependence of each of the outputs on the input signal values. So, each change in the value of the input signals immediately changes the state of the output signals. Unlike combinational logic circuits, the state of the outputs for the sequential logic circuit depends on the state of the input signals and the previous state of the output signals.

Figure 1.7 represents an example of the sequential logic circuit. As shown, the sequential circuit always includes feedback. Therefore,

Inputs   **Outputs**

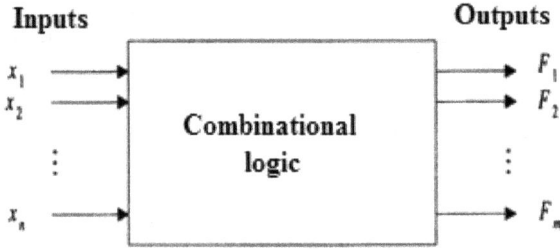

Fig. 1.6.   A general definition of the combinational logic circuit.

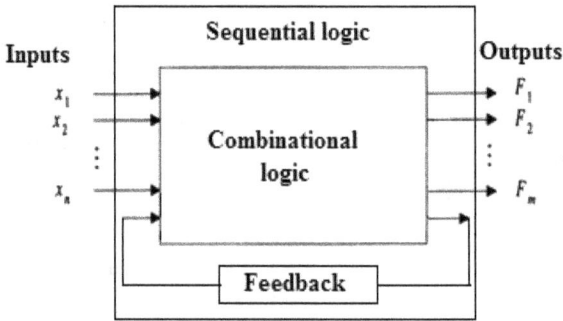

Fig. 1.7.   A general definition of the sequential logic circuit.

the sequential circuit always consists of feedback elements and a combinational environment.

The sequential logic circuit state depends on the previous state of the circuit; therefore, we can determine the order of the changes in the states for this circuit. This order of changes may depend on the clock. In this case such a change is called a synchronous behavior. If the system can change its state without relying on the clock, under influence of an occasional signal, such a system is called asynchronous.

As mentioned above, the logic states of the signal may be represented by logic "0" or logic "1" which characterize the low or high values of the signal. Usually, the behavior of the signal may be illustrated using a graphic form called an **impulse** or a **pulse**. In discrete electronics, a signal representing a **pulse** is defined as "a wave that departs from an initial level for a limited

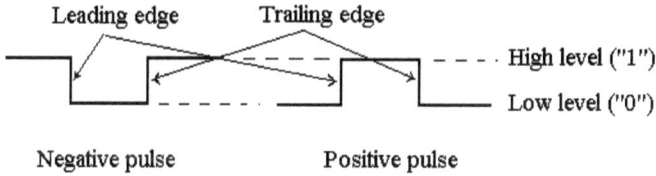

Leading edge   Trailing edge

High level ("1")
Low level ("0")

Negative pulse   Positive pulse

Fig. 1.8. The ideal pulse definition.

duration of time and ultimately returns to the original level." If we relate a pulse to the wave that exists between the two levels representing the two logic states, for example a pulse of voltage, we can show the pulse as positive or negative according to the initial state of the voltage. Figure 1.8 illustrates our definition of the pulses. Each pulse has a leading edge and a trailing edge. Evidently, the pulses shown in this picture are ideal pulses. In nature, processes which can change their states instantly, in such an ideal way, do not exist.

As shown, a pulse has two levels only: 0 and 1. The pulses may be of two different types: negative and positive. The type of a pulse is defined by the initial state. A logic system receives such pulses in its input and changes the output state according to the signals obtained. A number of such pulses and interval between them may be different. The pulses shown in Fig. 1.8 are named single pulse waveforms. Usually, the pulses come to the logic gate input in the structure of pulse train waveform. Figure 1.9 presents two different types of pulse train waveforms.

The pulse trains waveforms may be **periodic** and **aperiodic**. The periodic pulse train waveform may be characterized by a period T (a time after which the wave repeats itself) and a pulse duration t_p. A ratio of these two parameters appears in a novel general parameter called the **duty cycle**, which is defined by the following equation:

$$DC = \frac{T_p}{T} 100\%. \qquad (1.1)$$

The pulses shown in Figs. 1.8 and 1.9 are ideal pulses with abrupt edges. However, if we obtain such a pulse and feed it through the logic system, this system will disfigure the pulse as shown in Fig. 1.10.

Periodic pulse train

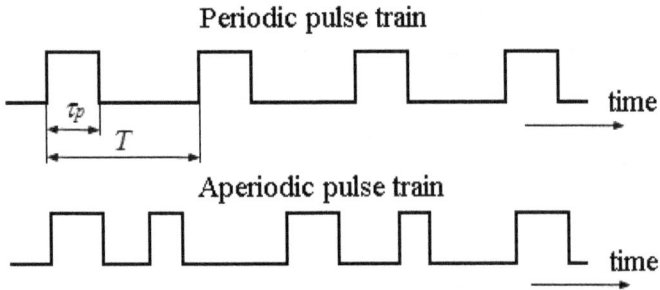

Aperiodic pulse train

Fig. 1.9.   Pulse train waveforms.

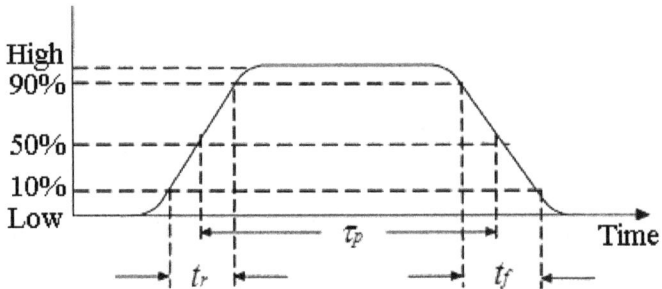

Fig. 1.10.   A non-ideal pulse.

The pulse shown in Fig. 1.10 differs from an ideal pulse in that it does not have abrupt edges. Instead, it has a finite **rise time** t_r (a time when the pulse level increases from 10% up to 90% of maximal value) and a finite **fall time** t_f (a time when the pulse level decreases from 90% down to 10% of maximal value). The 10% bands from the lower and upper levels provide stability boundaries of a logic system and their insensitivity to noise. These disfigurations are defined by real parameters of logic systems. Also, the response impulse of the logic system does not appear at the same moment that the input impulse was obtained. There is a certain delay between input and output, which is conditioned by the logic system structure and fabrication technology.

Figure 1.11 shows the pulse propagation through an inverting logic gate. Let us assume that at the initial time the input of our inverter is at the low logic state. Due to the general function of the

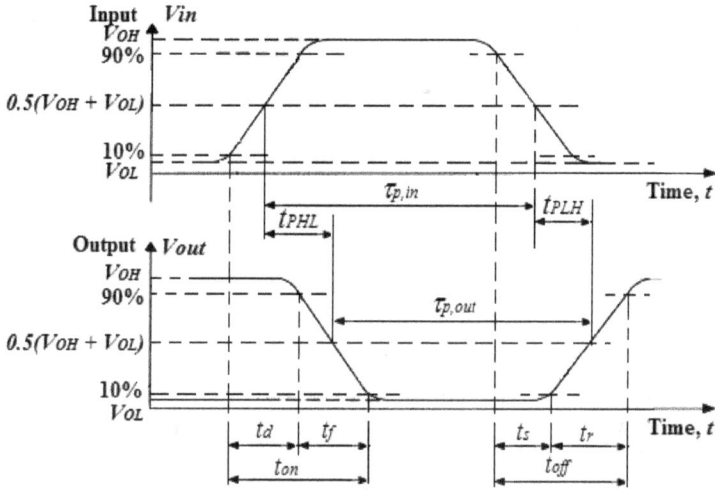

Fig. 1.11. Input-output waveforms for pulse propagation through a logic system.

inverter, its output state will be in the high state. We can say that the **system** or **logic gate is in the high logic state** when its output is high. Our diagram consists of two parts: upper and lower. The upper part presents the behavior of the input pulse and the lower part shows the response pulse behavior.

Let us also assume that our pulses represent the voltage changes in the input, V_{in}, and output, V_{out}, of the inverter. Here, V_{OL} and V_{OH} are the minimal and maximum voltages enabled for logic gates of the same logic family. In this diagram (Fig. 1.11), $\tau_{p,in}$ is the input pulse duration, $\tau_{p,out}$ is the output pulse duration, t_{PLH} and t_{PHL} are the time propagation delays which occur while increasing and decreasing the pulse respectively, and V_{OH} and V_{OL} are the maximum and minimum possible voltages in the input or output of the logic gate. If we assume that the input pulse duration is zero, the **propagation delay time** required by the pulse to pass through an inverter will be equal to $t_p = (t_{PHL} + t_{PLH})/2 = t_{PHL}/2 + t_{PLH}/2$. Thus, we estimate the time propagation delays t_{PHL} and t_{PLH} for decreasing and increasing pulses from half of the input pulse maximum value up to half of the output pulse maximum value only.

We can see and measure the finite **propagation delay time** t_d before the response begins. A **switching-on time** $t_{on} = t_d + t_f$ defines the appearance of the output impulse as shown in the figure. The **storage time** t_s and the **switching-off time** t_{off} may be defined like this: $t_{off} = t_s + t_r$. The output impulse is not equal to the input impulse generally. It may be reached if $t_d = t_s$ and $t_r = t_f$ only. To satisfy these conditions, our logic system should be symmetric and should use ideal inverters. Unfortunately, we can only come close to the ideal result, but not reach it. All parameters of pulses are significant for the proper design of logical systems and devices.

As shown in Fig. 1.11, the input and output voltages are functions of the time. We can represent these functions as follows:

$$V_{in} = \varphi(t) \quad \text{and} \quad V_{out} = \phi(t). \tag{1.2}$$

If we solve the first equation as $t = \psi(V_{in})$ and substitute this solution into the second equation, we obtain the following significant function called **the transfer function**:

$$V_{out} = f(V_{in}). \tag{1.3}$$

For the inverter, this function may be presented as shown in Fig. 1.12.

As shown in Fig. 1.12, the output voltage is high when the input voltage is low and the output voltage decreases as the input voltage

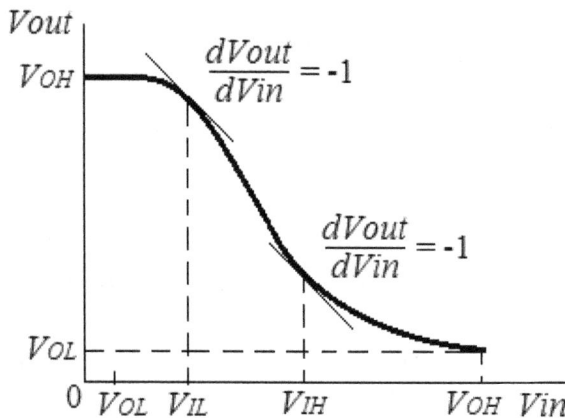

Fig. 1.12.   A transfer function view for the inverter.

rises. We can say that the logic gate is in the high state while the input voltage is in the interval of V_{OL}–$\boldsymbol{V_{IL}}$. This high state we designate as the logic "1". Also, we define the system state as a low state while the input voltage is in the interval of $\boldsymbol{V_{IH}}$–V_{OH}. This low state we designate as the logic "0". Therefore, our inverter may be in three states: the high state, the low state, and the transient or amplifying process. This transient process occurs for the input voltage interval of V_{IL}–V_{IH}, at the transition from one state to another. So, we can define the V_{IL} voltage as the maximum input voltage saving the system (logic gate) in the high state, "1". Also, the V_{IH} will be the minimal input voltage saving the system in the low state, "0".

Usually, the high and low states are stable. All variations of the input voltage in the ranges of V_{OL}–V_{IL} or V_{IH}–V_{OH} do not change the state of the system. These ranges are called the safety zones or the noise margins:

$$NM_L = V_{IL} - V_{OL} \quad \text{and} \quad NM_H = V_{OH} - V_{IH}. \qquad (1.4)$$

These noise margins are significant parameters of each logic system and they designate the stability level of the system. In the ideal case of a symmetrical system, these parameters are related as follows: $N_{ML} = N_{MH} = 0.5(V_{OH} - V_{OL})$. In accord with this relation, the transfer function for the ideal inverter will look like Fig. 1.13.

1.2. Basic logic families

When we talk about the various logic families, we mean the different technological bases used for the logic gates and systems fabrication. The diagram in Fig. 1.14 represents basic logic families and the technologies used. There are two basic semiconductor technologies: bipolar devices and unipolar devices. These technologies have emerged and evolved over the past several decades.

Logic systems began from the mechanical switching devices called relay. A principal electrical scheme and a photo of the relay are shown in Fig. 1.15. Such a system consists of an electromagnet and switching

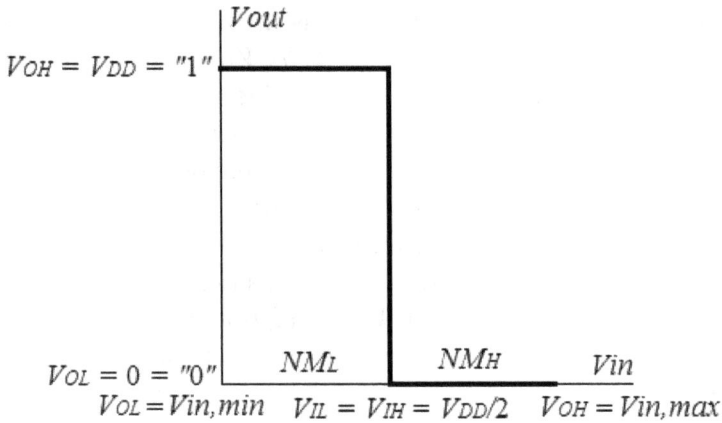

Fig. 1.13.   The transfer function of the ideal inverter.

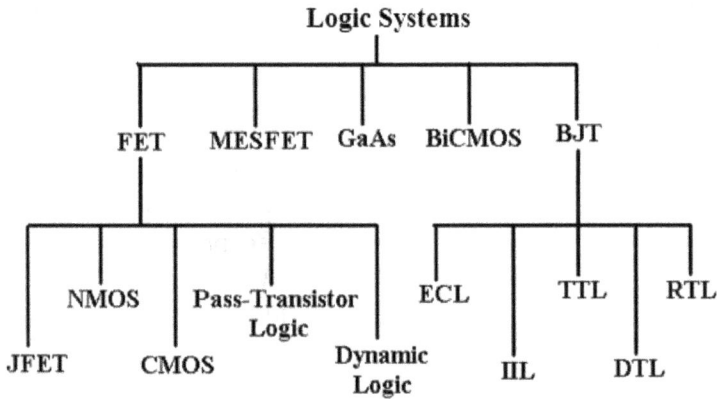

Fig. 1.14.   Basic logic families and technologies.

contacts. One of them is a normally open (NO) contact and another is a normally closed (NC) contact.

Here, there are two circuits which are not electrically connected: the control circuit consisting of the electromagnet and the operating circuit which has two different contacts, normally open and normally closed contacts. Evidently, the application of voltage to the controlling electromagnet can switch the contacts and realize the logical operation.

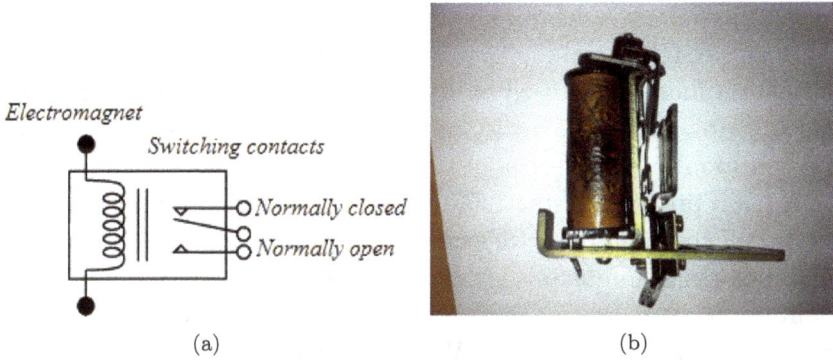

Fig. 1.15. Relay: (a) principal electrical scheme and (b) external view.

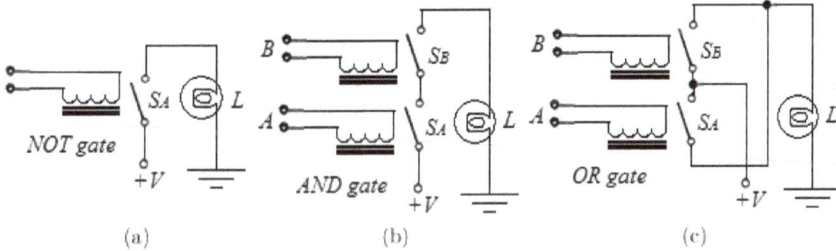

Fig. 1.16. Basic logical operations: (a) NOT, (b) OR, and (C) AND.

As is known, a complete consistent logical system can be represented by Boolean algebra. To realize the Boolean algebra, we need to provide three basic logic operations: inversion, logic addition, and logic multiplication. All these operations can be implemented using relay circuits as shown in Fig. 1.16. So, using relay logic gates, we can build a logic system of any complexity.

Logic systems built on the base of electro-mechanical relays represent ideal logic systems with ideal transform functions, however these systems suffer from several imperfections:

- Large dimensions;
- Low speed of switching;
- High electrical energy consumption;
- Presence of moving parts.

All these drawbacks become serious problems if we want to make the logic system smaller, faster, and more economical. Therefore, our efforts must be focused on finding other technological platforms which enable us to reach the small sizes, high speed and low power consumption.

A main property of the relay is its non-linear behavior. Thus, we need to choose non-linear devices when creating logic gates. First of all, bipolar semiconductor devices were used to achieve this goal. The main property of bipolar devices is their use of two types of charged carriers, electrons and holes. When the particles move, they have to overcome the p-n junctions. The bipolar semiconductor devices applicable to digital circuit technologies are as follows: diodes and bipolar transistors. The following logic systems, shown in Fig. 1.14, are built on the base of these devices:

- Diode-resistor logic, DRL;
- Resistor-transistor logic, RTL;
- Diode-transistor logic, DTL;
- Transistor-transistor logic, TTL;
- Emitter-coupled logic, ECL;
- Integrated-injection logic, I^2L.

Unlike the bipolar devices, the unipolar semiconductor devices work based on the transport of charged carriers of the same type, electrons or holes. Moreover, these particles do not cross the semiconductor junction during movement. Therefore, the noise level in unipolar devices is significantly lower than in bipolar devices. Unfortunately, the unipolar devices were invented later than the bipolar devices and technology evolved for them. To design digital circuits on the base of unipolar devices, we apply the following systems:

- Junction Field-Effect Transistor logic, JFET;
- N-type metal-oxide-semiconductor logic, NMOS;
- Complementary metal-oxide-semiconductor logic, CMOS.

CMOS technology has allowed the use of a unique system for the creation and optimization of logic elements and circuits. This system, in turn, led to the creation of pass-transition logic (PTL) and dynamical logic. Evidently, each logic family has their advantages and disadvantages. Therefore, the development of digital circuit technology has gone in several novel directions:

- the use of new materials for transistor fabrication, for example GaAs transistors;
- the use of novel principles for transistor fabrication, for example MESFET transistors;
- the integration of the principles of different logic families in one device, for example BiCMOS transistors;
- the transition from two-dimensional to three-dimensional transistor structures, and the application of nano-technology to prepare the digital logic devices.

Chapter 2

Logic Families Based on the Bipolar Devices

2.1. Diode logic

2.1.1. *The physical model of a junction diode*

A **diode** is a two-terminal electronic component with asymmetric conductance; it has low resistance to current flow in one direction, and high resistance in the other. The simplest semiconductor non-linear device is a junction diode. It consists of two contacting parts of the same semiconductor (a homo-junction device) doped by different impurities, donors and acceptors, providing suitable conductivity types, p- and n-types. Sometimes, to create the diode, these two parts are built from different semiconductor materials (a hetero-junction device). Figure 2.1 presents various types of diodes.

The part of the diode doped by p-type impurities is called the anode and the other part, which is doped by n-type impurities, is called the cathode. Due to the different concentrations of impurities in the two parts of the diode, the natural diffusion process occurs in the junction. Mobile positively charged carriers, or holes, begin moving from the p-type part into the n-type part and negatively charged carriers, or electrons, begin moving in the opposite direction. This process takes place as long as the diffused charged carriers create an internal, built-in, electrical field which prevents further diffusion.

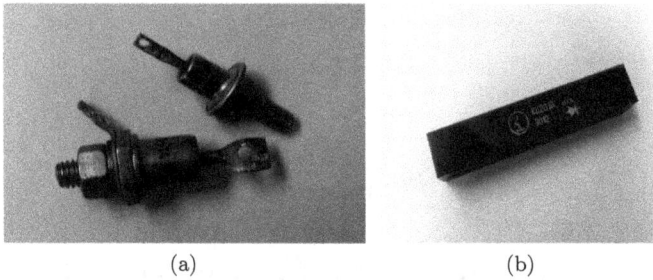

(a)   (b)

Fig. 2.1.   Various diodes: (a) silicon diodes and (b) a high voltage germanium diode.

Thus, the system of charges reaches a thermodynamic equilibrium or steady state.

Figure 2.2 illustrates a schematic representation of the junction diode. The left part of the device shown in Fig. 2.2(a), doped by acceptors, is the anode and the right part, doped by donors, is the cathode. An internal part with the width W is the depletion zone free of mobile charged carriers. The anode and cathode, outside of the depletion zone, represent a quasi-neutral semiconductor with balanced charged carriers of both types. Figure 2.2(b) represents the charge distribution inside the junction diode.

As shown, only stationary ion-charges remain in the depletion region. This way, a high-strength electrical field is built-in inside the depletion region and compensates for the external electrical field produced by the quasi-neutral anode and cathode. A voltage generated by this field, a built-in potential, prevents the flow of current through the diode at low forward voltage. If the applied forward voltage is more than the built-in potential, a current begins to flow through the diode. Under the applied voltage, the width of the depletion region decreases and the resistance of the diode is defined by a semiconductor structure only. The diode resistance becomes minimal and constant. If external voltage is applied to the diode, we can measure the current driving through it and build the current-voltage (I–V) characteristic illustrating the process. Figure 2.3 represents various approximation models describing diode behavior under applied voltage. These models show the I–V characteristics of the diode given different assumptions.

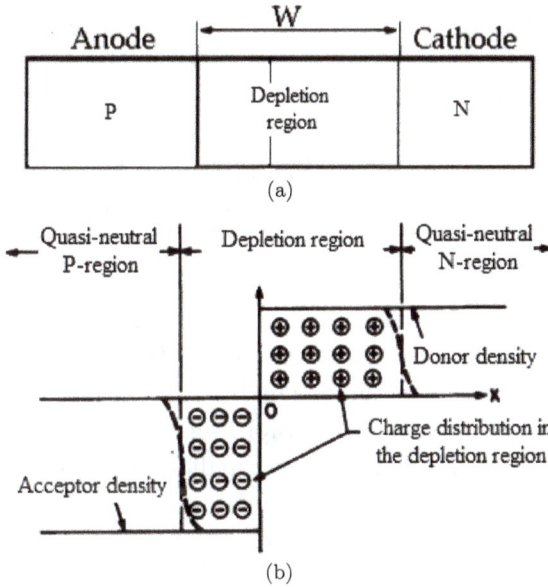

Fig. 2.2. Schematic view of the diode: (a) section of the diode and (b) charge distribution in the diode.

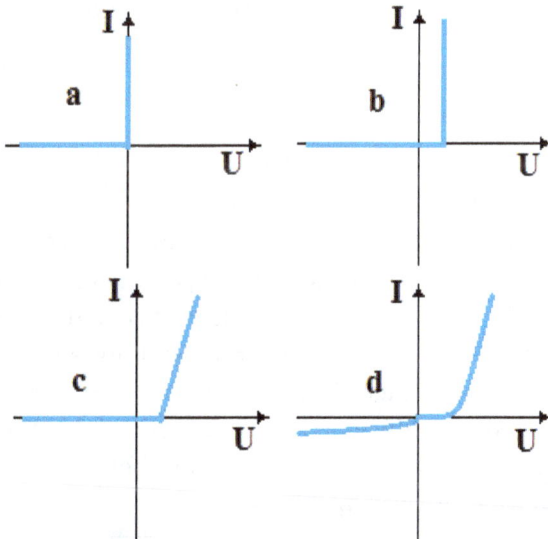

Fig. 2.3. I–V characteristics of diodes: (a) an ideal diode, (b) a diode with a rectifying threshold, (c) a diode with finite internal resistance and (d) the I–V characteristic described by Shockley's model.

The principle of operation for diodes is based on a general understanding of solid state physics. Here we present only a short description of diodes and models showing their behavior in the simplest circuits. Figure 2.3 represents a diode as a device with non-linear behavior. As shown in Fig. 2.3(a), the main goal of this device is to transfer a current through the diode under the positive voltage applied to the diode and to close all current in the reverse direction. Figure 2.3(b) shows a first approximation of a diode; this model explains that a diode has an internal electrical field which characterizes it. Figure 2.3(c) illustrates the presence of a finite internal resistance in the diode. The best approximation for an ideal diode is the Shockley equation graphically presented in Fig. 2.3(d). This well-known equation describes the diode behavior in both direct and reverse applied voltage. This equation is as follows:

$$I = I_s \left(e^{\frac{V_a}{V_t}} - 1 \right). \tag{2.1}$$

Where I is the current through a diode, V_a is the voltage applied to the diode (it may be as positive as well as negative), $V_t = kT/q$ is a thermal voltage characterizing the environmental temperature, k is the Boltzmann constant, T is the environmental temperature measured in K and q is the elementary charge. I_s is a leakage current or a reverse saturation current which may be described by the following equation:

$$I_s = qAn_i^2 \left(\frac{D_n}{L_n N_D} + \frac{D_p}{L_p N_A} \right). \tag{2.2}$$

Where A is the diode area, n_i is an intrinsic concentration of free charged carriers in the semiconductor, N_D and N_A are the impurity concentrations in the cathode and anode parts of a diode respectively, D_n and D_p are the diffusivity of electrons and holes respectively, and L_n and L_p are the average diffusion lengths of electrons and holes respectively. Here, a diffusion length is defined by the lifetime of charged carriers, which is the average time between the generation and recombination of these particles:

$$L_n = \sqrt{D_n \tau_n}, \quad L_p = \sqrt{D_p \tau_p}. \tag{2.3}$$

Where τ_n and τ_p are the lifetimes of electrons and holes respectively.

2.1.2. *The equivalent scheme of a diode*

To measure the current flows through a diode, the diode must be connected with the supply which can vary the applied voltage. Figure 2.4 represents a measuring electrical circuit and the measured characteristic.

As shown in Fig. 2.4(b), this I–V characteristic looks significantly non-linear. We can select two virtual points characterizing the diode: the voltage at the beginning of the current flow through the diode, $V_{D,\gamma}$, and the voltage associated with the maximum current flow, $V_{D,on}$. For silicon diodes, these voltages are approximately 0.5 V and 0.7 V respectively. The point of $V_{D,\gamma}$ characterizes a threshold of the current beginning due to decreasing of the internal dynamical resistance, r_γ, from infinite to finite. The point of $V_{D,on}$ characterizes the minimum constant internal resistance of the diode. Between these two points, the internal resistance of the diode dynamically changes. By this way, we can call three areas limited by a defined voltage drop on the diode as follows:

(a) $V_D < V_{D,\gamma}$ — the cutoff area;
(b) $V_{D,\gamma} < V_D < V_{D,on}$ — the dynamical area;
(c) $V_{D,on} < V_D$ — saturation area.

Figure 2.5 illustrates these distinguished points with the example of water poured from the pot through the hole. The lower section of

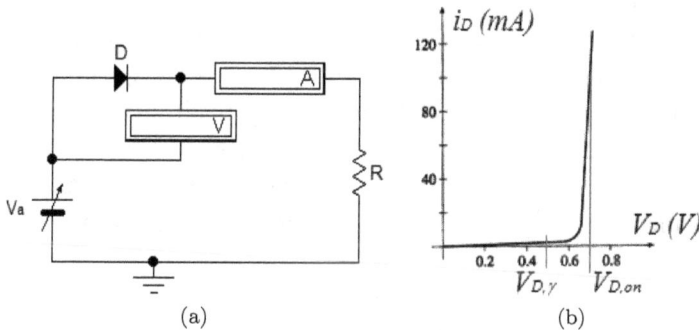

Fig. 2.4. Some applications of a diode: (a) Measuring the electrical circuit and (b) measuring the I–V characteristic.

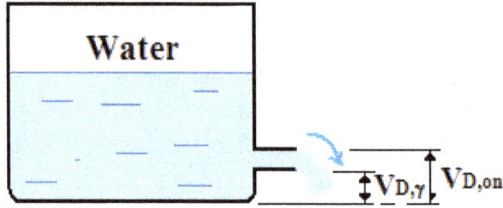

Fig. 2.5.   An illustration of the work of the diode.

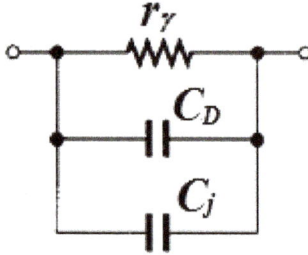

Fig. 2.6.   An equivalent circuit of the diode.

the drain hole, $V_{D,\gamma}$, is a threshold that characterizes the beginning of flow of water. If the water level is below the threshold, the water cannot flow naturally. If the water level is upward of the upper section of the hole, $V_{D,on}$, the water stream will be constant. Between these two extreme points, a flow is variable.

Using the concept of internal dynamic resistance, we can build an equivalent circuit of the diode. Figure 2.6 presents an equivalent circuit for a diode.

This equivalent circuit consists of three elements which together define the behavior of the diode under voltage. C_D represents the diffusion capacitance of the diode. This capacitance is determined by the amount of charge transferred through the diode due to the positive voltage applied between the anode and cathode. This capacitance may be calculated according to the following equation:

$$C_D = \frac{dQ}{dV} = \frac{d(i_D t)}{dV}. \tag{2.4}$$

Where Q is a transferred charge, i_D is the forward current through a diode, and t is the time. Evidently, this capacitance is at its

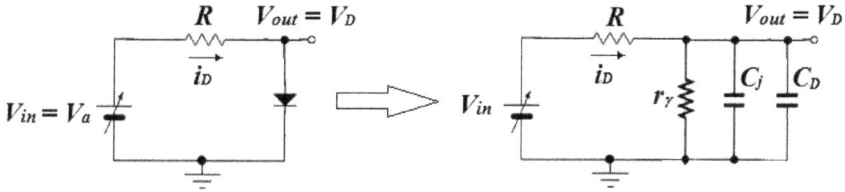

Fig. 2.7. An equivalent circuit of a diode in the electrical system.

minimum at an applied voltage lower than $V_{D,\gamma}$. This capacitance, C_D, increases with a forward current growth.

C_j represents the junction capacitance of the diode. This capacitance is determined by the thickness of the depletion zone in the diode junction. This capacitance may be calculated according to the following equation:

$$C_j = \frac{\varepsilon_0 \varepsilon_r A}{W} = A\sqrt{\frac{q\varepsilon_0\varepsilon_r N_A N_D}{2(N_A + N_D)(V_b - V_a)}}. \qquad (2.5)$$

Where ε_0 is the permittivity of vacuum, ε_r is the relative permittivity of the applied semiconductor, and V_b is the built-in potential of the junction. A built-in potential may be estimated using the following equation:

$$V_b = V_t ln\frac{N_A N_D}{n_i^2}. \qquad (2.6)$$

If voltage exceeding the $V_{D,on}$ is applied to the diode, the depletion region disappears and consequently the junction capacity also disappears. Under reverse voltage, the junction capacitance decreases up to a minimum defined by the length of the diode.

So, to analyze diode behavior in the electrical circuit, the diode can be replaced with an equivalent scheme, see Fig. 2.6, as shown in Fig. 2.7.

Figure 2.8 represents an analysis of diode behavior in a switching circuit. Part (a) shows time-diagrams of the input voltage, the current behavior under various input voltages, the charge state as a function of time, and the voltage on the diode. At the initial time, $t < t_1$, the voltage across the diode is missing. At the point t_1, the

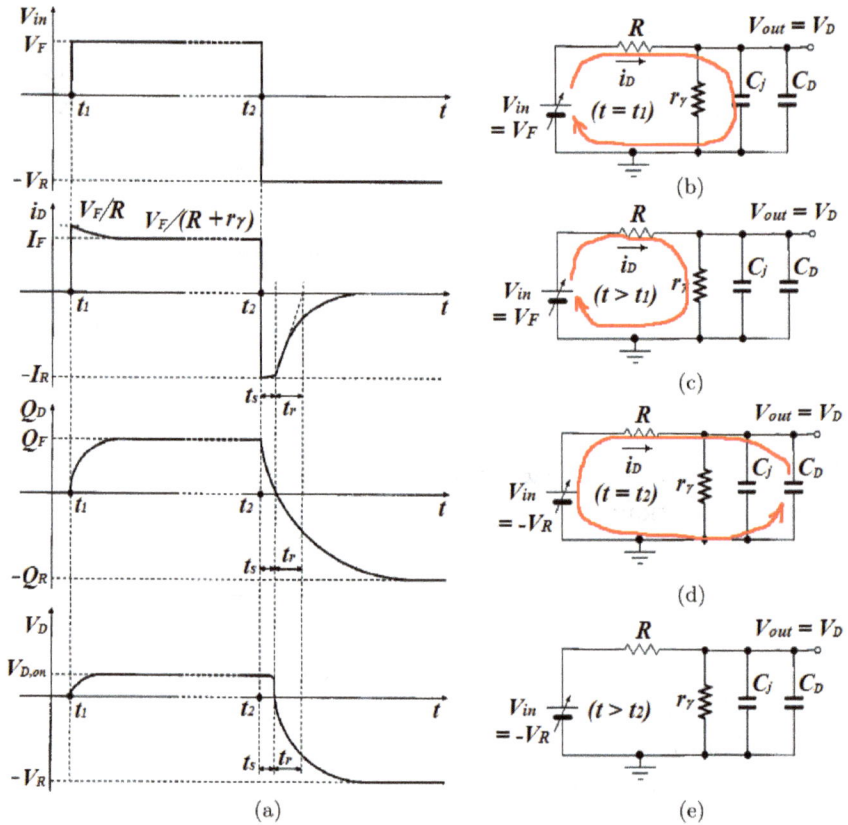

Fig. 2.8.   Diode behavior in a switching circuit analysis.

input voltage quickly increases to the value $V_F > V_{D,on}$. At this point, the internal dynamic resistance is infinite, the diffusion capacity is very small due to the missing of the charge transfer, and the junction capacity represents zero resistance due to an abrupt increase in the input voltage. Figure 2.8(b) illustrates the current behavior at this time. As shown in the current diagram, the first-time current value is equal to V_F/R. After that, the junction capacitance disappears, the diffusion capacitance is charged, and the value of the internal resistance is reduced to a minimum. Figure 2.8(c) presents the final state of the current in the circuit, $i_D = V_F/(R + r_\gamma)$. A diagram of charge, Q_D, illustrates the charge behavior in the time interval

$t_1 - t_2$. The diffusion capacity is charged up to the charge Q_F through a definite time $(\tau = C_D R)$ and remains constant. The voltage on the diode increases up to $V_{D,on}$ through the same time and remains constant such that this voltage is defined by the minimum internal resistance of the diode. This state remains up to time t_2.

At the point t_2, an input voltage abruptly changes from $V_{Fon} - V_R$. At this time the charge accumulated in the diffusion capacity is discharged through a resistor R up to zero and disappears; this capacity exists at a forward voltage only. The time of this discharge is called t_s, a storage time. Through this time, a constant reverse current flows in the circuit, as shown in Fig. 2.8(d). After that, the junction capacity appears and is charged up to value of $-Q_R$. The time required for this process may be estimated through a timing constant t_r (a recovery time), which is equal to $t_r = C_j R$. Here, a value of C_j is defined by the length of the diode. In other words, the length of the depletion region, W, rises up to the maximum, i.e. the length of the diode. Figure 2.8(e) represents the final state of the circuit: all currents are zero (we neglect the leakage current), the junction capacity is charged up to $-Q_R$, and the maximum negative (reverse) voltage, $-V_R$, is applied to the diode.

2.1.3. *Analysis of a diode's switching circuit behavior*

If we are going to analyze the switching circuit, our tasks usually include the identification of the logic implemented in the circuit, the determination of all currents and voltages in the circuit, the building of the transfer function of the circuit, and sometimes the performance of various additional tasks. To solve the first part of the tasks, an identification of the type of logic, we usually apply the building of a truth-table. This is the simplest and most direct analytical method. The second part, an estimation of currents and voltages, enables us to calculate power dissipation in the circuit and sometimes the value of the maximum fan-out. The third part, a transfer function, enables us to evaluate the circuit in comparison with other schemes and technologies and to estimate the value of the noise margins, NM_L and NM_H, and the propagation delay for the circuit.

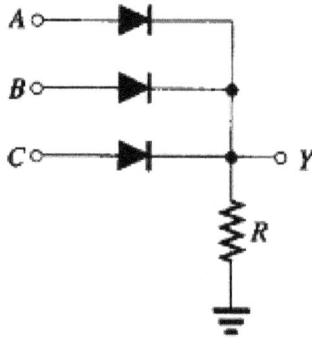

Fig. 2.9.   A diode switching scheme.

For example, let us consider the circuit presented in Fig. 2.9 which relates to Problem 1.

Example 1.
It is known that $V_A = 5$ V, $V_B = 4.8$ V, $V_C = 0.7$ V, $R = 4$ kΩ, $V_{D,on} = 0.7$ V, and $V_{D,\gamma} = 0.5$ V.

1. Identify the logic type of the circuit.
2. Estimate the minimum and maximum output voltage (logic "0" and "1").
3. Evaluate the state of each diode for the input voltages given.
4. Build a transfer function.

Solution 1.

1. As the names for all the input points to the scheme (A, B, C) are chosen at random, we can consider them equal, that is, this circuit is symmetric with respect to inputs. If one of the applied input voltages exceeds the value $V_{D,on}$, an output voltage V_y will repeat the input voltage. Therefore, this circuit realizes the "**OR**" logic.
2. To solve the second part of the problem, we need to choose the input that will transfer a maximum voltage to the output. If we choose a voltage applied to the input B, $V_B = 4.8$ V, to achieve this goal, we can conclude that V_y will be equal to:

$$V_y = V_B - V_{D,on} = 4.8 - 0.7 = \underline{4.1 \text{ V}}.$$

However, in this case the voltage drop on the diode A is $V_A - V_y = 5 - 4.1 = 0.9$ V, and that is impossible as the voltage drop on the diode cannot be more than $V_{D,on}$. Therefore, our choice was wrong. To obtain the output voltage, we must choose the maximum input voltage from parallel connected inputs. The right solution is as follows:

$$V_y = V_A - V_{D,on} = 5 - 0.7 = \underline{\mathbf{4.3 \ V}}.$$

Therefore, the logic "1" is equal to 4.3 V and the logic "0" is equal to 0 V.

3. Now we can evaluate the state of each diode for the input voltages given:

   Diode A: $V_A - V_y = 5 - 4.3 = 0.7$ V — saturation state.

   Diode B: $V_B - V_y = 4.8 - 4.3 = 0.5$ V — dynamical state.

   Diode C: $V_C - V_y = 0.7 - 4.3 = -3.6$ V — cutoff state.

4. To build a transfer function, we need to understand circuit behavior. To this end, we shorted inputs and obtained a circuit with one input. This enabled us to study circuit behavior when there is only one variable input. The input voltage varies from zero to the maximum voltage, 5 V in our case. The input diode begins to conduct a current, after which V_D will rise up to $V_{D,\gamma}$. Between voltages $V_{D,\gamma}$ and $V_{D,on}$, the dynamical resistance of the diode decreases hyperbolically and after $V_{D,on}$ it becomes constant. Therefore, we can present this behavior as shown in Fig. 2.10. The transfer characteristic is shown approximately, we have neglected the behavior of the output voltage in the interval $V_{D,\gamma} - V_{D,on}$.

   An analysis of the built transfer characteristic shows that the logic "1" in this circuit is the "poor 1". Several circuits connected in series can significantly decrease the logic "1" that may be not applicable in the digital system. So, we must change the circuit to increase the logic "1".

The following circuit, seen in Fig. 2.11, represents an improved version of the scheme shown in Fig. 2.9. This circuit was created specifically to improve the output voltage of this logic gate.

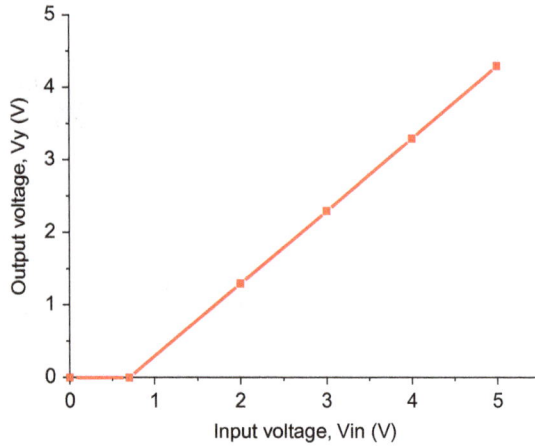

Fig. 2.10.   The transfer function for the circuit shown in Fig. 2.9.

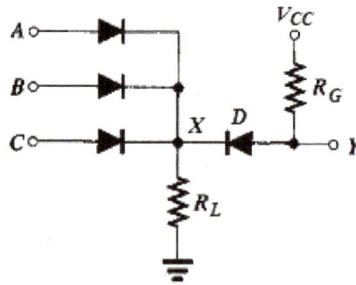

Fig. 2.11.   An improved diode switching scheme.

Example 2.

In the same conditions that characterized **Problem 1**:

1. Estimate the minimum and maximum output voltage (logic "0" and "1").
2. Build a transfer function.

Solution 2.

1. To solve the first part of the problem we must short all inputs and consider two cases:

$$V_A = V_B = V_C = V_{in} = V_{CC} = \text{"1"}$$

and

$$V_{in} = \text{``0''}.$$

If $V_{in} = \text{``1''}$, the voltage at point X, V_X, will be equal to $(V_{in} - V_{D,on})$ V, therefore

$$V_Y = V_X + V_{D,on} \approx V_{CC}.$$

If $V_{in} = \text{``0''}$, we must consider two partial circuits according to the Kirchhoff's circuit law: $V_{CC} = iR_G + V_{D,on} + iR_L$ and $V_{CC} = iR_G + V_Y$. Joint solution of these equations gives the following relation for the output voltage V_Y:

$$V_Y = \frac{V_{CC}R_L + V_{D,on}R_G}{R_L + R_G}$$

where $R_G \gg R_L$, therefore

$$V_Y \approx V_{D,on} + V_{CC}\frac{R_L}{R_G} \approx V_{D,on}.$$

Thus, the addition of a diode D and resistance R_G for the circuit enables us to increase the output voltage, logical "1", however the logical "0" is also changed.

2. Now, we can build the transfer characteristic of the circuit using the obtained results. This will be a straight line beginning approximately at the point with coordinates (0.7 V, 0.7 V). Figure 2.12 represents the I–V characteristic of the improved circuit. As shown, we now have the high value of the logical "1" and we paid for this improvement by increasing the logical "0".

2.1.4.   *The Schottky diode*

Each diode must be connected with other elements of the electrical circuits. These connections are made via metal contacts and wires. However, to decrease electrical losses across contact points, we need to consider these points and evaluate the processes occurring in the semiconductor-metal contact. Figure 2.13 depicts two possible types of metal-semiconductor contacts. The contact defined by a

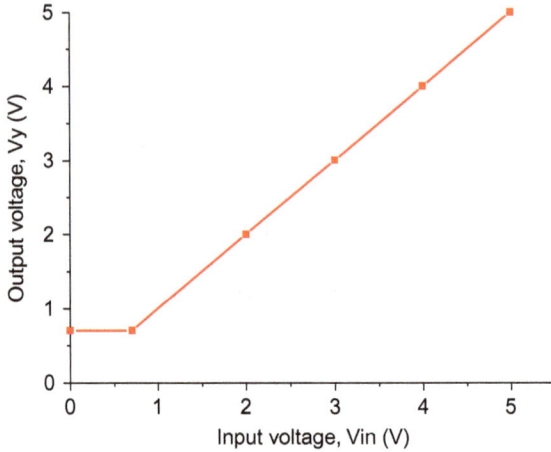

Fig. 2.12.　The transfer function for the circuit shown in Fig. 2.11.

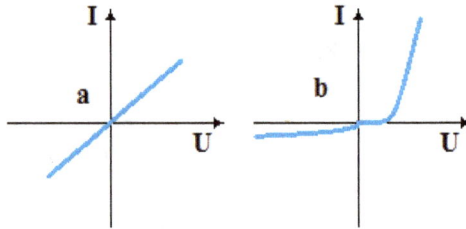

Fig. 2.13.　Two possible types of metal-semiconductor contacts: (a) Ohmic contact and (b) rectifying contact.

direct relation between the applied voltage and the measured current (Fig. 2.13(a)) and described by Ohm's law, $V = R \cdot I$, where R is the contact resistance, known as the Ohmic contact. The second contact has a significantly non-linear shape and is known as the rectifying contact. Its I–V characteristic looks like the junction diode characteristic. However, the physical processes occurring in this case are different from the processes taking place in the junction diodes.

The first and most important difference is that the metal and the semiconductor are two different materials. A metal may be characterized by the overlapping of the conductive band and the

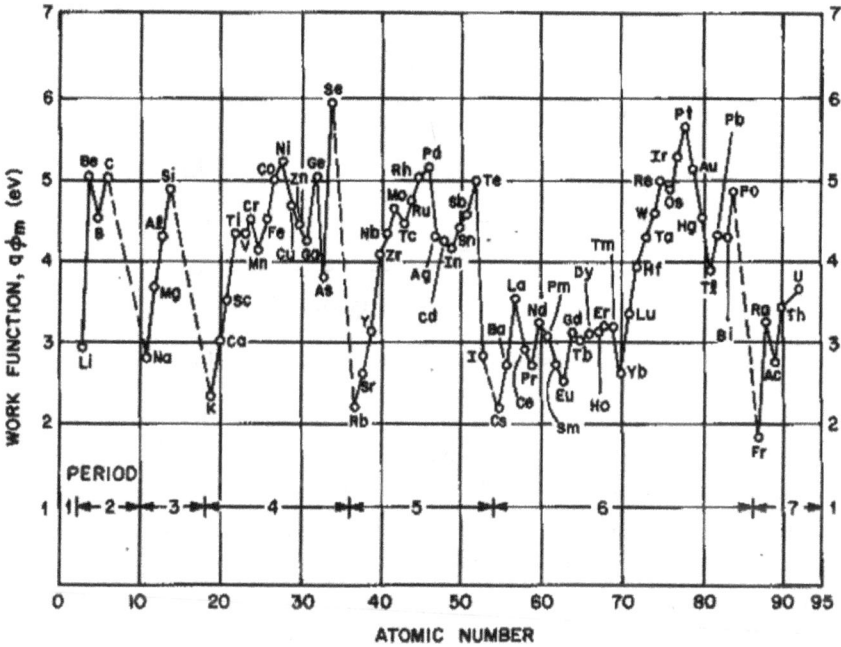

Fig. 2.14. The metal work functions for a clean metal surface in a vacuum versus an atomic number (after Michaelson, *IBM J. Res. Dev.*, 22, 1978, 78).

valence band. Thus, the conductive electrons are also the valence electrons. The free conductive electrons that belong to each atom constituting the metal crystal lattice fasten the atoms together. This type of connection is referred to as a metallic type. In this case, the Fermi level is within the overlapping zone. To liberate an electron from the metal, it is necessary to apply an energy equal to the work function to the metal. Figure 2.14 presents the metal work functions for a clean metal surface in a vacuum versus an atomic number.

To liberate an electron from a semiconductor, we also need to apply an energy equal to the work function. However, the amount of energy needed to liberate an electron from a semiconductor is different from the amount of energy needed to liberate an electron from a metal. In the semiconductor, the conducting band and the valence band are non-overlapping. They are separated by the so-called band gap. This energy band does not trap electrons in the

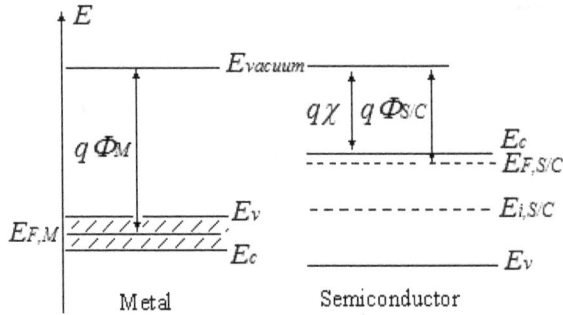

Fig. 2.15.   Schematic representation of metal and semiconductor in the energy axis.

ideal crystal lattice. Due to this, the connection between atoms in a semiconductor is only made at the expense of neighboring atoms. This type of connection is referred to as an atomic or a covalence type. Figure 2.15 schematically represents the definition of the work function for metals and semiconductors.

In this figure, two materials with different work functions are shown. For the metal, the work function is the energy value required to liberate an electron from the Fermi level outside the metal, $q\Phi_M$.

For the semiconductor, the work function is the sum of the conductive band width ($q\chi$, electron affinity) and the distance between the Fermi level and the conductive band's bottom, $q\Phi_{S/C} = q\chi + (E_c - E_{F,S/C})$. Here, E_c designates the bottom of the conductive band, E_v designates the top of the valence band, $E_{i,S/C}$ is the Fermi level for an intrinsic semiconductor and $E_{F,S/C}$ is the Fermi level for the doped semiconductor. As known, $E_{F,S/C}$ and $E_{i,S/C}$ are related by the following equation for the semiconductor of n-type:

$$E_{F,S/C} = E_{i,S/C} + kTln\frac{N_D}{n_i} = \frac{E_c - E_v}{2}$$

$$+ \frac{3}{4}kTln\left(\frac{m_h^*}{m_e^*}\right) + kTln\frac{N_D}{n_i} \qquad (2.7)$$

where m_e^* and m_h^* are the effective masses of an electron and a hole respectively. These effective masses are different for various crystallographic planes, therefore the values of these parameters may be different in various sources.

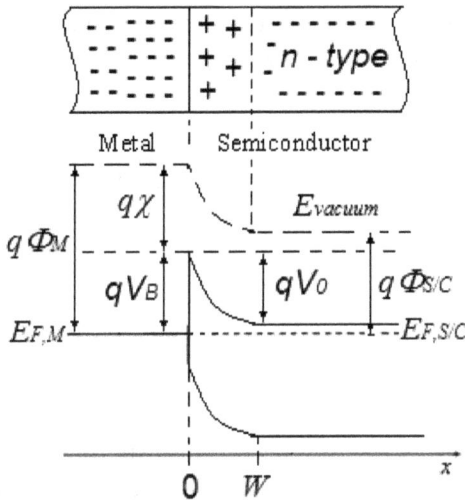

Fig. 2.16. The formation of a potential barrier on the interface between two media.

Figure 2.15 illustrates a semiconductor of n-type and a metal with a work function greater than the semiconductor's work function. If we put these two different materials in contact, the process of transition to the thermodynamical equlibrium immediately begins. Thus, the Fermi levels of both materials have to become equal. This leads to the bending of energy bands with the appearing of a potential barrier on the boundary between two materials. Figure 2.16 illustrates this process.

The free mobile electrons begin to cross the interface between the semiconductor and the metal due to the difference between the work functions of metals and semiconductors. They move from the semiconductor to the metal. In this way, a depletion region is formed in the frontier of the semiconductor. This process goes on until the appearance of the built-in electrical field limiting the emission of electrons from the semiconductor to the metal. This electric field arises due to charge separation. Thus, the process of the alignment of the Fermi levels reaches a steady state.

The width of the depleted zone, designated by the value W, may be estimated in accordance with the theory of the junction diodes as long as the difference in the concentration of impurities in the metal and the semiconductor is more than two orders. Thus, a value W may be estimated in the following way (see Eq. (2.8)):

$$W = \sqrt{\frac{2\varepsilon_0\varepsilon_r(V_0 - V_a)}{qN_D}}. \tag{2.8}$$

Where V_0 is the built-in potential, V_a is the applied external voltage, and V_B is the potential barrier height. Evidently, this value, W, behaves in the same way as the depletion zone width in the usual junction diodes.

In the case of contact between a semiconductor of n-type and a metal with a work function lower than in the semiconductor, the thermodynamical equalization process will go the other way. Now, the free mobile electrons begin to cross the interface in the reverse direction and we do not obtain a barrier in such an interface. This process is illustrated in Fig. 2.17.

We have considered two possible variants of metal-semiconductor contacts. However, there are two other possible variants. All four possible variants are presented in the form of a table (see Table 2.1), in which these variants are fully defined.

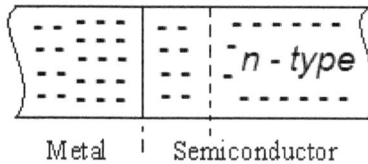

Fig. 2.17.   Ohmic contact in the border metal-semiconductor.

Table 2.1.   The types of metal-semiconductor contacts.

Type of s/c	$q\Phi_M$ or $q\Phi_{S/C}$	Type of contact
N	>	Blocking
N	<	Ohmic
P	>	Ohmic
P	<	Blocking

To describe an electrical transport through a blocking contact, we must take into account the mechanism of electron liberation. There are several mechanisms enabling the liberation of electrons from one material to another. Each of them depends on different conditions. All of the mechanisms act together, however, in each case, one of them dominates. The basic mechanisms are as follows: diffusion current, thermionic emission (Schottky mechanism), tunneling (Fawler–Nordheim mechanism), and emission from traps (Poole–Frenkel emission).

If the current through a metal-semiconductor contact is proportional to the square of the environmental temperature under direct voltage bias, such a current is called a thermionic current. It flows when the n-type semiconductor has a work function lower than in the metal (see Fig. 2.16). This current is described by the following equation:

$$I = AA^*T^2 e^{-\frac{V_B}{V_t}} \left(e^{\frac{V_a}{nV_t}} - 1 \right) = I_0 \left(e^{\frac{V_a}{nV_t}} - 1 \right). \qquad (2.9)$$

Where n is the ideality coefficient and A^* is Richardson's coefficient, which is equal to $A^* = \frac{4\pi q m^* k^2}{h^3} = [\frac{A}{m^2 k^2}]$. Such a contact is called the **Schottky diode**.

In the case of a reverse voltage applied to the contact, the termionic current disappears and a diffusion reverse current begins to flow. This current is described as follows:

$$I_s = \frac{q A N_D D_p}{L_n}. \qquad (2.10)$$

The potential barrier in the Schottky diode is significantly lower than in the junction diodes; it is equal to approximately 0.2–0.3 V. A comparative picture is shown in Fig. 2.18.

2.2. The bipolar junction transistor (BJT) and related logic families

2.2.1. *The BJT transistor — principles of operation*

A **transistor** or **triode** is a semiconductor device which is constructed of three semiconductor layers arranged consecutively and

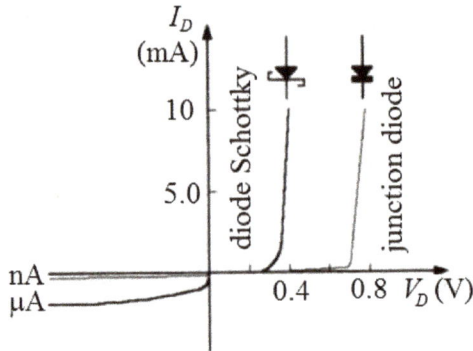

Fig. 2.18.   Volt-Amper characteristics of different diodes.

Fig. 2.19.   Transistors.

which has three terminals. Figure 2.19 represents various types of bipolar junction transistors.

There are two options when connecting the three layers: N-P-N and P-N-P. Therefore, there are two types of transistor: NPN and PNP. Figure 2.20 represents a semiconductor layer arrangement (Fig. 2.20(a)) and a schematic designation of the device (Fig. 2.20(b)).

The transistor NPN shown in Fig. 2.20 is a bipolar device such that charged carriers of both types which are moving through this device are crossing two junctions and must go through the semiconductor layers doped by the opposite type. A principle of

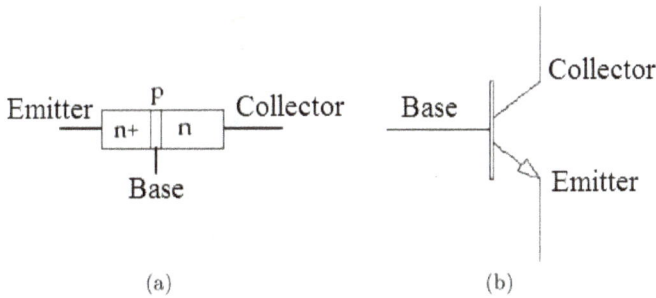

Fig. 2.20. Schematic illustration and schematic designation of the NPN transistor.

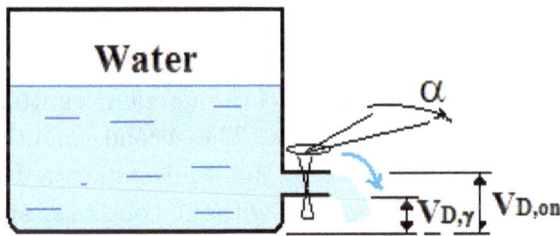

Fig. 2.21. Principle of operation for the semiconductor transistor.

operation of the transistor is illustrated by the picture shown in Fig. 2.21. A transistor may be considered a source of current which is controlled by a small variation of the angle in the flow valve. A small variation in this angle leads to a big change in the flow.

To fabricate the transistor, we must meet the three following conditions:

1. Each semiconductor layer must be doped with a different level of impurities, like this: emitter:base:collector $= 10^{19}:10^{17}:10^{15}$.
2. The thickness of the base must be lower than the mean free path of the minor charged carriers. In the case of the NPN transistor these will be electrons.
3. The interface area of the collector-base contact must be greater than the area of the base-emitter interface by 3–50 times.

Compliance with these conditions can simplify the development and estimation of currents flowing through the transistor. The first

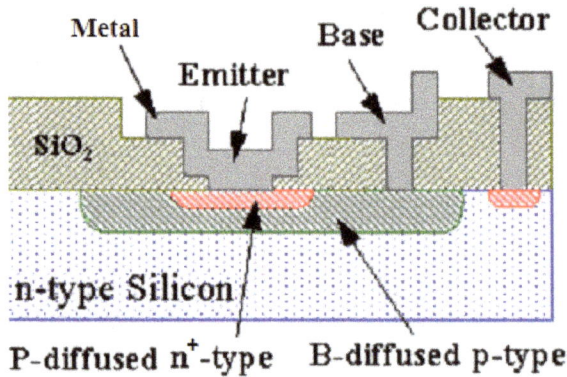

Fig. 2.22.   Bipolar transistor structure.

condition enables us to consider each junction, emitter-base, and base-collector as a one-sided diode. The second condition restricts the thickness of the base layer of the semiconductor. It allows the electrons to cross the base region without collisions, and therefore without recombination. The third condition ensures the asymmetry of the transistor and a more complete collection of electrons. A more accurate picture of the transistor's structure fabricated by planar technology, not drawn to scale, is given in the following Fig. 2.22.

As shown in Fig. 2.22, the transistor consists of three semiconductor layers: n^+, p, and n. A thin base layer of p-type has been arranged between the emitter and collector layers. The interface between collector and base is significantly larger than the base-emitter interface. A very highly doped region of the n^+ type has also been arranged between the collector and its electrode to create an Ohmic type contact with the collector.

So the NPN transistor includes three layers with electrodes, and we can connect these electrodes with two external supplies in various ways. Each connection will bias a suitable semiconductor junction towards a direct or reverse voltage. These voltages on the junction define four various modes for transistor behavior. These modes, according to the definition, are shown in Table 2.2.

Here, EBJ designates the emitter-base junction and CBJ designates the collector base junction. As in the case of a diode, we have

Table 2.2. The modes of transistor behavior.

EBJ	CBJ	Operation mode
Reverse	Reverse	Cutoff
Forward	Reverse	Active
Forward	Forward	Saturation
Reverse	Forward	Reverse active

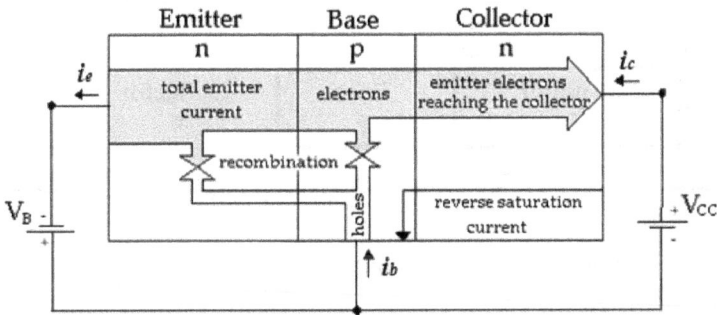

Fig. 2.23. The operation of an NPN transistor in the active mode.

two junctions of PN types connected by the anodes. This structure defines several models of a transistor using the properties of diodes. The schematic presentation of a transistor presented in Fig. 2.20(a) may be shown in expanded view to illustrate transistor behavior in the active mode.

As shown in Fig. 2.23, a voltage from the supply V_B is applied to the junction BE in the direct direction and a voltage V_{CC} is applied to the junction BC in the reverse direction. The currents flowing in the circuit, I_E (emitter current), I_C (collector current), and I_B (base current) are also shown in this figure. Due to Kirchhoff's law, these currents form a closed loop, therefore

$$i_B + i_C = i_E. \tag{2.11}$$

Let us consider how the transistor works. Due to the direct bias of the BE junction, electrons injected from the V_B supply to the emitter cross it and go to the base. As the thickness of the base is

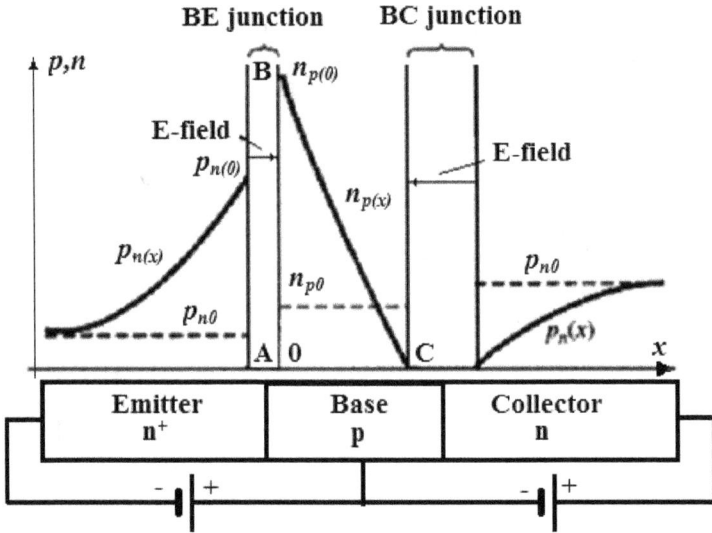

Fig. 2.24. The distribution of minority charged carriers within a transistor in the active mode.

lower than that of the mean free path of the electrons, most of them move through the base without collisions and enter the collector. As the collector is inversely biased, the electrons drift to the collector electrode and close the circuit created by the V_{CC} supply. A small number of electrons recombines in the base. A schematic distribution of minority charged carriers within the transistor shown in Fig. 2.24 helps us to understand the transistor's behavior.

Parameters designating the impurity concentrations in different parts of the transistor are shown as n — the concentration of electrons — and p — the concentration of holes. So, the majority charged carriers' concentration in the n-type part is designated as n_n and the majority charged carriers' concentration in the p-type part is p_p. According to this definition, parameters p_n and n_p are designated the minority charged carriers' concentrations. An index 0, when appended to the concentration, designates the concentration without voltage: p_{n0}, n_{p0}. According to the "Law of Junctions,"

$$p_n(x) = p_{n0}exp(V/V_t). \tag{2.12}$$

Therefore, a concentration of minority charged carriers on the border of the emitter under voltage V_{BE}, applied from the V_B supply, will be equal to

$$n_p(0) = n_{p0}\exp(V_{BE}/V_t) \qquad (2.13)$$

where V_{BE} is the direct voltage on the base-emitter junction.

An electron current injected from the emitter to the base will be proportional to the junction area, the charge of the electrons, the diffusivity of minority carriers (electrons) in the base and the concentration change in the base:

$$I_n = A_E q D_n \frac{dn_p(x)}{dx} \qquad (2.14)$$

where A_E is the area of the BE junction and $n_p(x)$ is the electron distribution along the base. As shown in Fig. 2.24, the electron distribution along the base may be approximately depicted by a straight line, thus the Eq. (2.14) will transform as follows:

$$I_n = A_E q D_n \left(-\frac{n_p(0)}{w}\right) = -\frac{A_E q D_n}{w} n_{p0} e^{\frac{V_{BE}}{V_t}} . \qquad (2.15)$$

Most of the electrons which have reached the base are jumping at the collector without collisions since the base width is shorter than the mean free path of electrons. Therefore, we can write that **the collector current is equal to the electron current**, $i_c = -I_n$. Due to the law of charge conservation, $n_{p0} = n_i^2/N_A$, so

$$i_c = \frac{A_E q D_n n_i^2}{w N_A} e^{\frac{V_{BE}}{V_t}} = I_s e^{\frac{V_{BE}}{V_t}} . \qquad (2.16)$$

Where I_s is the leakage current, its value is usually of $I_s = 10^{-12} - 10^{-15}$ A, depending on the transistor's dimensions. As I_s is proportional to the square of the intrinsic concentration for the given semiconductor and it is known that

$$n_i^2 = RT^3 e^{\frac{E_g}{kT}} . \qquad (2.17)$$

Where R is the coefficient of proportionality, the leakage current is very dependent on the temperature and approximately doubles for every rise in temperature of $5°C$.

It is interesting to note that the collector current is independent of the voltage across the base-collector (BC) junction. In other words, until the collector voltage rises above the voltage at the base, practically all electrons passing the base are collected by the collector.

As shown in Fig. 2.23, a base current consists of two components: the first of them is a current comprised of holes injected to the emitter and the second one is the recombination current comprised of the electrons which entered the base and cannot overcome it without collisions. The first component may be calculated according to Eq. (2.16), with changes made to allow for minority carriers in the emitter and so that it is proportional to the exponent of the voltage applied to the BE junction:

$$i_{b1} = \frac{A_E q D_p n_i^2}{N_D L_p} e^{\frac{V_{BE}}{V_t}}. \tag{2.18}$$

The second component is proportional to the charge entered into the base, Q_n, related to the life-time of electrons as minority carriers in the base $i_{b2} = Q_n/\tau_b$. This charge is equal to the area under the line BC in the triangle ABC (see Fig. 2.24) multiplied by the BE-junction area:

$$Q_n = A_E q \frac{1}{2} n_p(0) w. \tag{2.19}$$

Substituting the value of $n_p(0)$ from Eq. (2.13), we obtain for i_{b2} the following equation:

$$i_{b2} = \frac{1}{2} \frac{A_E q w n_i^2}{N_A \tau_b} e^{\frac{V_{BE}}{V_t}}. \tag{2.20}$$

So, the base current will be equal to

$$i_b = i_{b1} + i_{b2} = \left(\frac{A_E q D_p n_i^2}{N_D L_p} + \frac{1}{2} \frac{A_E q w n_i^2}{N_A \tau_b} \right) e^{\frac{V_{BE}}{V_t}}$$

$$= \frac{I_s}{\beta} e^{\frac{V_{BE}}{V_t}} = \frac{i_c}{\beta}. \tag{2.21}$$

Where β is known as **an amplification coefficient for the circuit with a common emitter:**

$$\beta = \frac{1}{\frac{D_p N_A w}{D_n N_D L_p} + \frac{1}{2} \frac{w^2}{D_n \tau_b}}. \tag{2.22}$$

As can be seen from Eq. (2.22), the magnification coefficient β is a constant value which depends on the chosen material (D_p, D_n, τ_b), the concentrations of impurities (N_A, N_D) representing an applied technology, and the width of the base, w. This coefficient is usually equal to 5–200. Combining together Eqs. (2.11) and (2.22), we can designate one other coefficient called **the amplification coefficient in the circuit with the common base, α,** as follows:

$$\alpha = \frac{\beta}{\beta + 1}. \tag{2.23}$$

All the above considerations apply to the analysis of the transistor in the active state. However, there is the reverse active state also, as shown in Table 2.2. To separate the parameters related to the two different modes of operation of the transistor, we introduce the notation F (forward) and R (reverse) in the designation of coefficients α and β. So, we have the β_F and β_R coefficients and the α_F and α_R coefficients, related by Eq. (2.24). The reverse coefficients are significantly lower than the forward coefficients. This is due to differences in the impurity concentrations in the emitter and collector and differences at the junction areas.

For convenience, we can define several other coefficients which can be used as references. All these coefficients are related to the coefficient β and help to analyze it. The parameter γ_E, called **the emitter efficiency**, is an approximation of Eq. (2.22) for the defined value of the base thickness:

$$\gamma_E = \frac{1}{1 + \frac{D_p N_A w}{D_n N_D L_p}}. \tag{2.24}$$

$$\alpha_T = 1 - \frac{t_r}{\tau_b} = 1 - \frac{w^2}{2 D_n \tau_b}. \tag{2.25}$$

This parameter is called **the base transport factor**, which takes into account the time needed for electrons to cross the base. Both these parameters are related to the magnification coefficient α:

$$\alpha = \gamma_E \alpha_T. \tag{2.26}$$

Figure 2.25 represents the possible connections of transistors in electrical circuits.

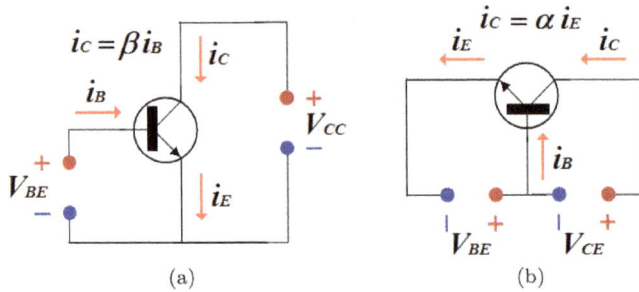

Fig. 2.25.   Different schemes of connections for transistors in electrical circuits:
(a) a circuit with a common emitter and (b) a circuit with a common base.

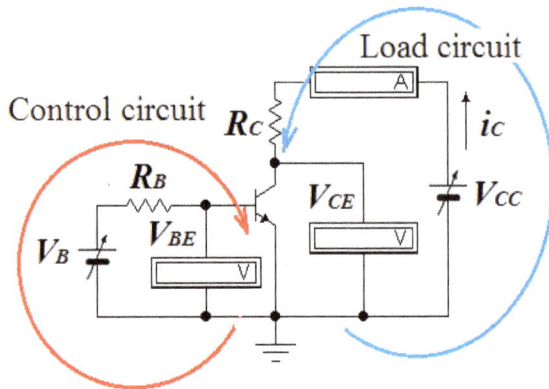

Fig. 2.26.   Typical connection of the transistor with a common emitter.

Typically, the common emitter connection is used in amplifying and digital circuits. Such a circuit, equipped with measuring devices and resistors, is presented in Fig. 2.26. Also, two different circuits are shown in the figure. We have here the control circuits, marked by red, which control the base current through the transistor, and the load or power circuit, marked by blue, supplied from the external power supply and controlled by the control circuit. The variable control and power supplies enable us to provide measurement of the voltages V_{BE} and V_{CE} and the collector current in the circuit and to build the output or load characteristics of the transistor.

These output characteristics with approximate values of transistor parameters are presented in Fig. 2.27. As shown, the I–V

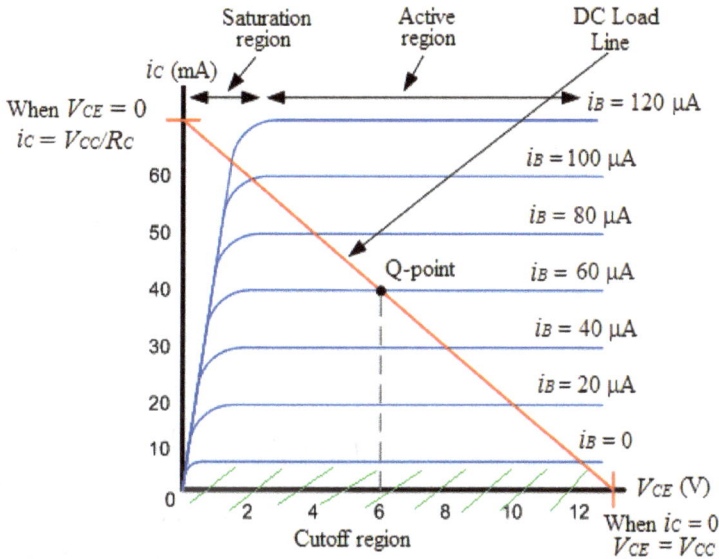

Fig. 2.27. Output characteristics of the NPN transistor.

characteristics family of the transistor represents a series of like dependencies varied with the applied base current. Each current curve begins as a straight line and flexes when the transistor enters a saturation state with the growth of the V_{CE} voltage. We can see three working regions in this picture: the saturation region, the active region, and the cutoff region, marked by green.

The red line connecting the two extreme points on the graph is called a work-line. These points are virtual as a collector's current cannot reach zero and cannot reach the value of V_{CC}/R_C. The point Q is the usual working point in the linear transistor amplifier. The collector's current is in active mode in linear or amplification circuits. In digital circuits, this current usually takes the values of cutoff or saturation.

2.2.2. The model of Ebers–Moll and modes of operation

As shown in Fig. 2.20, the transistor consists of two PN junctions or, virtually, of two diodes connected face-to-face (anode-to-anode).

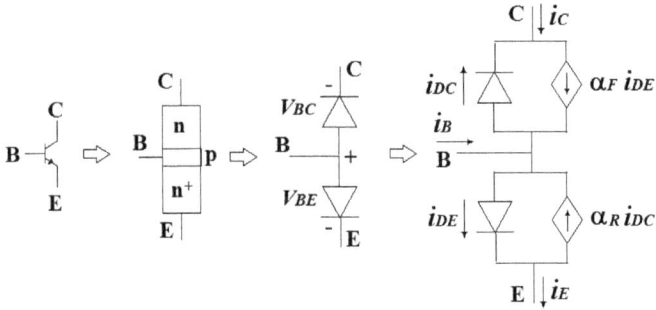

Fig. 2.28.   The model of Ebers–Moll for the NPN transistor.

However, for the diode we have the very good model described by the
Shockley equation or Eq. (2.1). Therefore, we can present a current
through a diode as a current defined by two power sources: one
of them is the controlled supply V_{BE} or V_{BC} and the second one
represents a reverse saturation current. Using such a representation
of a transistor, we can present it according to the model shown in
Fig. 2.28.

In this model, we replaced the PN junctions on the diodes
with additional current generators. Therefore, the currents through
the collector and emitter diodes, i_{DC} and i_{DE}, will be expressed
according to Eqs. (2.27) and (2.28).

$$i_{DE} = I_{sE}\left(e^{\frac{V_{BE}}{V_t}} - 1\right) \tag{2.27}$$

$$i_{DC} = I_{sC}\left(e^{\frac{V_{BC}}{V_t}} - 1\right) \tag{2.28}$$

where I_{sE} and I_{sC} are reverse saturation currents for both diodes.
It was stated in the previous section that the areas of the junctions
are not equal and because of this, the collector saturation current,
I_{sC}, is more than the I_{sE}. Therefore, one can determine the following
relation between these currents and the coefficients α_R and α_F:

$$\alpha_R I_{sC} = \alpha_F I_{sE} = I_s. \tag{2.29}$$

Now, using Fig. 2.26 we can add three other equations to
Eqs. (2.27) and (2.28). As the transistor represents a node, we can

use Kirchhoff's law to write the following equations:

$$i_E = i_{DE} - \alpha_R i_{DC}. \tag{2.30}$$

$$i_C = -i_{DC} + \alpha_F i_{DE}. \tag{2.31}$$

$$i_B = i_E - i_C = i_{DE}(1 - \alpha_F) + i_{DC}(1 - \alpha_R). \tag{2.32}$$

The joint solution of Eqs. (2.27), (2.28), (2.30), (2.31), and (2.32) allows us to obtain a mathematical model of Ebers–Moll which enables us to analyze the transistor behavior in all its operation modes.

$$\begin{cases} i_E = \dfrac{I_s}{\alpha_F}\left(e^{\frac{V_{BE}}{V_t}} - 1\right) - I_s\left(e^{\frac{V_{BC}}{V_t}} - 1\right) \\[2mm] i_C = I_s\left(e^{\frac{V_{BE}}{V_t}} - 1\right) - \dfrac{I_s}{\alpha_R}\left(e^{\frac{V_{BC}}{V_t}} - 1\right) \\[2mm] i_B = \dfrac{I_s}{\beta_F}\left(e^{\frac{V_{BE}}{V_t}} - 1\right) + \dfrac{I_s}{\beta_R}\left(e^{\frac{V_{BC}}{V_t}} - 1\right) \end{cases} \tag{2.33}$$

As an example, let us consider the use of the Ebers–Moll model to study the transistor behavior in various operation modes.

Example 3.
Using the model of Ebers–Moll, show the different modes of operation of the transistor NPN.

Solution 3.

1. **An active mode.**
   In this state, the applied supplies are as follows: $V_{BE} > 0$ and $V_{BC} < 0$.
   Therefore, $e^{\frac{V_{BC}}{V_t}} \ll 1$ and $I_{DC} \approx -I_{sC}$ and our Eq. (2.33) will transform into the following equations:

$$\begin{cases} I_C = I_s\left(e^{\frac{V_{BE}}{V_t}} - 1\right) - \dfrac{I_s}{\alpha_R} \\[2mm] i_E = \dfrac{I_s}{\alpha_F}\left(e^{\frac{V_{BE}}{V_t}} - 1\right) - I_s \end{cases}.$$

2. A cutoff mode.

In this state, we have $V_{BE} < 0$ and $V_{BC} < 0$. Therefore, $e^{\frac{V_{BE}}{V_t}} \ll 1$ and $e^{\frac{V_{BC}}{V_t}} \ll 1$.

Thus, $I_{DC} \approx -I_{sC}$ and $I_{DE} \approx -I_{sE}$ and our system transforms as follows:

$$\begin{cases} I_C = I_s - \dfrac{I_s}{\alpha_R} \\ i_E = \dfrac{I_s}{\alpha_F} - I_s \end{cases}.$$

3. A saturation mode.

Now, $V_{BE} > 0$ and $V_{BC} > 0$. Equation (2.33) remains the same:

$$\begin{cases} i_C = I_s\left(e^{\frac{V_{BE}}{V_t}} - 1\right) - \dfrac{I_s}{\alpha_R}\left(e^{\frac{V_{BC}}{V_t}} - 1\right) \\ i_E = \dfrac{I_s}{\alpha_F}\left(e^{\frac{V_{BE}}{V_t}} - 1\right) - I_s\left(e^{\frac{V_{BC}}{V_t}} - 1\right) \end{cases}.$$

In the case of $V_{BE} \gg V_t$ and $V_{BC} \gg V_t$, these equations transform as follows:

$$\begin{cases} i_C = I_s e^{\frac{V_{BE}}{V_t}} - \dfrac{I_s}{\alpha_R} e^{\frac{V_{BC}}{V_t}} \\ i_E = \dfrac{I_s}{\alpha_F} e^{\frac{V_{BE}}{V_t}} - I_s e^{\frac{V_{BC}}{V_t}} \end{cases}.$$

2.3. The digital inverter and the resistor-transistor logic (RTL) family

Figure 2.29 presents a simple inverter based on the relay.

As shown in Fig. 2.29, this circuit consists of two different independent supplies (a control supply, V_{in}, and a power supply, V_S), a load resistor (R_L), and a non-linear switching element implemented by the relay (*SW*). We have two circuits related through an electromagnetic field. If we designate a logic "0" equal to 0 V and a logic "1" equal to 5 V, our system will work as an inverter. The main deficiencies of such inverter are long response time, large size, and high power consumption. To improve this circuit, we can replace the

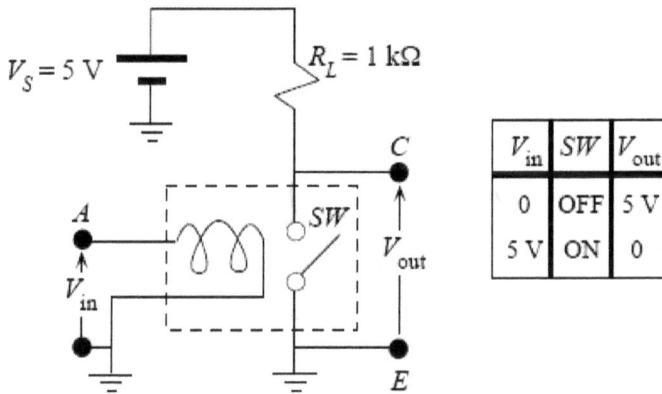

Fig. 2.29.   The simplest digital inverter.

V_{in}	SW	V_{out}
0	OFF	5 V
5 V	ON	0

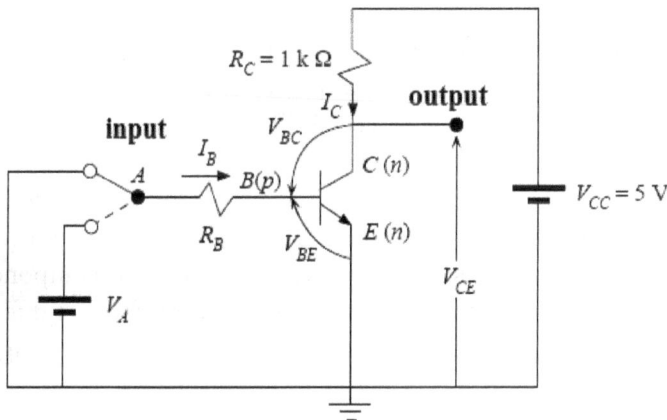

Fig. 2.30.   The simplest transistor inverter.

non-linear element relay with another non-linear element: we replace the relay with the transistor as shown in Fig. 2.30. Now the relation between the control and power circuits is performed using a transistor and the transfer charged carriers through it.

The circuit presented in Fig. 2.30 consists of two related parts: an input circuit and an output circuit. As shown, an input voltage can take only two values, designated as logical "0" and logical "1". In positive logic, "0" is the lower value, zero, and "1" is the higher value of the supply V_A. A power circuit is supplied by the power supply V_{CC}, independent of the input supply. This circuit works as follows:

1. $V_{in} = 0$, therefore the transistor is in the cutoff mode, the collector current is approximately zero and the output voltage is equal to the V_{CC} or "1".
2. $V_{in} = V_A$, we choose an input voltage so that the transistor enters into saturation. In this mode, it transfers a maximum collector current and practically all voltage V_{CC} drops across the resistor R_C; the output voltage will be a minimum equal to the $V_{CE,sat}$ which is equal to the silicon transistors, or approximately 0.1–0.2 V only.

So, we obtained an inverter based on the BJT transistor. Using the components of this circuit: resistors and transistors, such a circuit is called a resistor-transistor logic (RTL) circuit.

Figure 2.31 represents a typical RTL inverter. Its I–V characteristics are shown in Fig. 2.32. This circuit may be characterized using the load line in the I–V relations and the transfer function. As is known, the load line is a geometrical solution of the system presented in the circuit and represents the behavior of all the components of the circuit. In our system, the load line is a straight line due to the linear behavior of a current through the resistor Rc, according to the output equation:

$$V_{out} = V_{CC} - i_C R_C. \tag{2.34}$$

At the same time, the current i_C flows through the transistor and relates with the input voltage, as shown by the Ebers–Moll model, for example.

We can find this relation using the Ebers–Moll model. For our circuit (Fig. 2.31), the system of equations will look as follows:

$$\begin{cases} V_{CC} = i_C R_C + V_{CE} & \textit{output equation} \\ V_{in} = i_B R_b + V_{BE} & \textit{input equation} \\ V_{CE} = V_{BE} - V_{BC} & \textit{coupling equation} \end{cases} \tag{2.35}$$

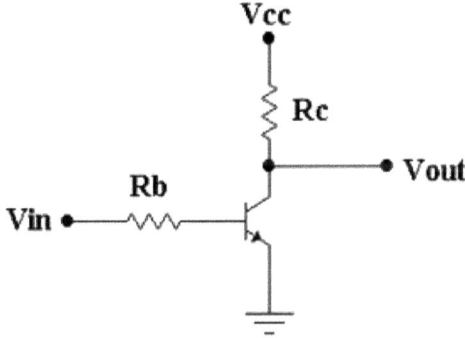

Fig. 2.31. A typical RTL inverter.

Taking into account Eq. (2.29), we can transform the collector current's equation from the Eqs. (2.33) in the following way:

$$i_C = \alpha_F I_{ES} \left(e^{\frac{V_{BE}}{V_t}} - 1 \right) - I_{CS} \left(e^{\frac{V_{BC}}{V_t}} - 1 \right) = \alpha_F I_{ES} \left(e^{\frac{V_{BE}}{V_t}} - 1 \right)$$

$$-I_{CS} \left(e^{\frac{V_{BE}-V_{CE}}{V_t}} - 1 \right) = \alpha_F I_{ES} e^{\frac{V_{BE}}{V_t}} - \alpha_F I_{ES}$$

$$-I_{CS} e^{\frac{V_{BE}}{V_t}} e^{-\frac{V_{CE}}{V_t}} + I_{CS} = \left(\alpha_F I_{ES} - I_{CS} e^{-\frac{V_{CE}}{V_t}} \right) e^{\frac{V_{BE}}{V_t}}$$

$$-(\alpha_F I_{ES} - I_{CS}). \tag{2.36}$$

From this equation we obtain

$$e^{\frac{V_{BE}}{V_t}} = \frac{i_C + \alpha_F I_{ES} - I_{CS}}{\alpha_F I_{ES} - I_{CS} e^{-\frac{V_{CE}}{V_t}}}. \tag{2.37}$$

Therefore, we obtain a relation between an input voltage and the collector current:

$$V_{BE} = V_t ln \frac{i_C + \alpha_F I_{ES} - I_{CS}}{\alpha_F I_{ES} - I_{CS} e^{-\frac{V_{CE}}{V_t}}}. \tag{2.38}$$

Assuming $I_{CS} \approx 0$, we obtain the following applicable approximation:

$$V_{BE} = V_t ln \left(\frac{i_C}{I_S} + 1 \right) \approx V_t ln \left(\frac{i_C}{\alpha_F I_{ES}} \right). \tag{2.39}$$

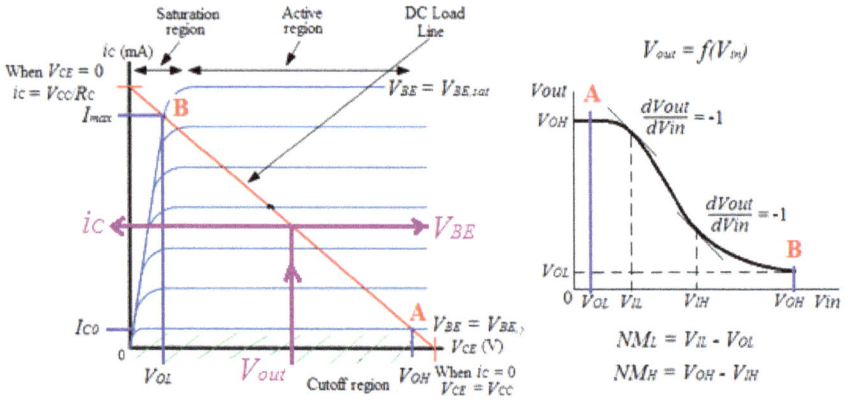

Fig. 2.32.   The I–V characteristics and the transfer function of the RTL circuit.

So, the coupling equation relates the input and output parameters of our circuit. Now we can create the transfer function using Eq. (2.39).

The two extreme points designated in Fig. 2.32 as A and B represent two modes of work for the transistor: the cutoff state (A) and the saturation state (B). A digital inverter, while working in a steady state, may be in only one of these states. The load line between these two extreme states is the transition from one stable state to another. These points exist in the transfer function diagram also. The transition between stable states is an amplification mode not applicable to digital RTL circuits.

To build a transfer function, we need to take a point from the load line representing an output voltage $V_{out} = V_{CE}$ (marked by violet). Using the output Eq. (2.34), we can calculate the collector current (i_C), and using the coupling Eq. (2.39), we can obtain the voltage falling on the BE-junction, V_{BE}. After that, using the input Eq. (2.35), we can calculate the input voltage relating to the chosen output voltage and designate the obtained point with coordinates (V_{in}, V_{out}) in the diagram of the transfer function. Performing these operations many times and combining all of the calculated points, we get the graph of our transfer function.

This method is straight, simple, and precise; however, it is very time-consuming and it does not provide information on the effect of the various parameters of the circuit on its behavior. Therefore, to

$$V_{out} = f(V_{in})$$

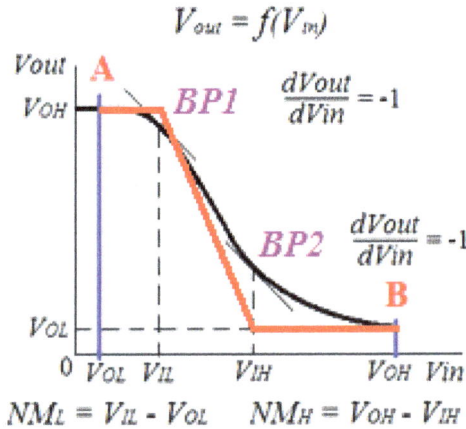

Fig. 2.33.   A piecewise approximation of the transfer function.

build a transfer function, we use a less accurate, approximate method of building, called the **piecewise** method. The basic idea behind this method is to calculate the coordinates of a few essential points of the transfer function that determine the appropriate behavior of the circuit. These points on the graph we connect using straight-line pieces, so that the behavior of the circuit between adjacent points is approximately described by linear equations. An example of this approach is presented in Fig. 2.33. As shown here, the transfer function is replaced by its piecewise approximation (marked by red).

Two essential points, called breakpoints (**BP**) in Fig. 2.33, are the transition points for the transistor's behavior. They mark the transition from the cutoff mode to the linear (amplification) mode, *BP1*, and the transition from the linear mode to the saturation mode, *BP2*. Evidently, the transition from the cutoff mode to the linear mode is a continuous process. However, we approximately appear this process as a non-continuous within the one defined point: in this point the collector current through the transistor is equal to 1% of the maximum possible current, $i_{C1} = 0.01 I_{max} = 0.01 V_{CC}/R_C$. So, the first transition point, *BP1*, is defined. The second point, the point at which there is a transition from the linear mode to the saturation mode or vice versa, we can define using the basic relation of the amplification mode (2.21), $i_C = \beta \cdot i_b$. In the saturation mode, this relation is not right, $i_C < \beta \cdot i_b$. So, we define the second

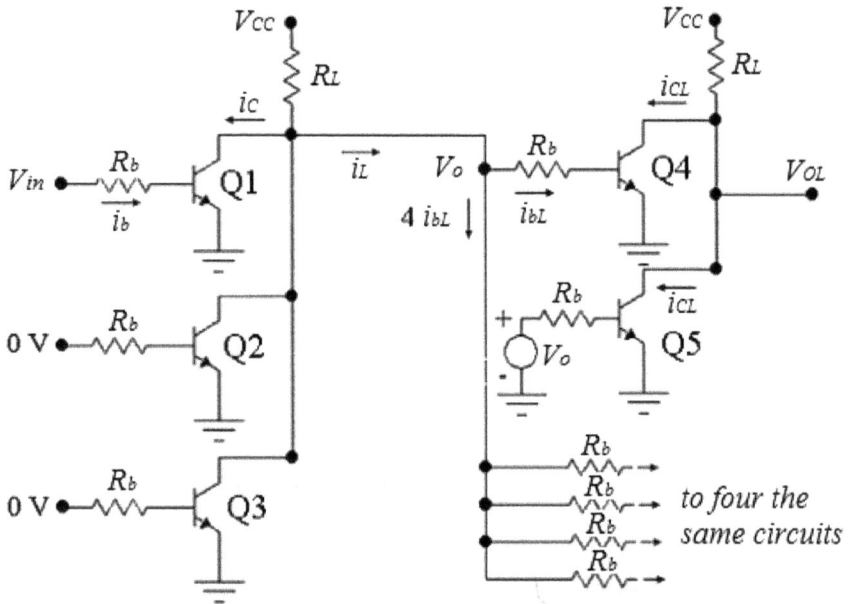

Fig. 2.34.   Typical RTL circuit.

transition point, *BP2*, as the point at which Eq. (2.21) begins to be right. Now, we have two break points, BP1 and BP2, for our circuit (see Fig. 2.31) which define the transition points of the transistor behavior for different input voltages. These points we connect with straight lines, which define the behavior of the circuit between break points. Moreover, now we can calculate the noise margins for our circuit as shown in Fig. 2.33 and compare the transfer function of our circuit with transfer function of the ideal inverter (see Fig. 1.13).

As an example, consider the typical RTL circuit. Let us build the transfer function $V_{out} = f(V_{in})$ for this circuit.

Example 4.
Figure 2.34 represents a complex RTL circuit. Build the transfer function $V_o = f(V_{in})$ for this circuit.

Solution 4.

To build the transfer function, first of all we need to clearly understand the circuit's behavior. In this circuit we see three identical

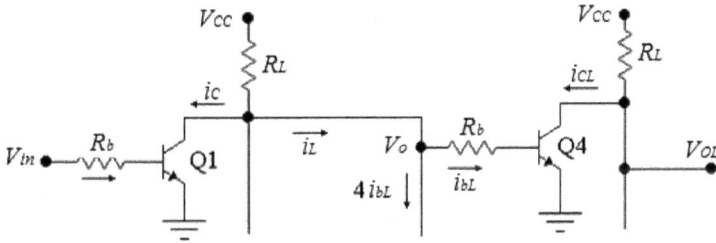

Fig. 2.35. The simplified circuit.

input branches connected together which have the same collector resistor. So, we can write that the fan-in of our circuit is equal to $m = 3$. A load part of the circuit connected to the output voltage V_o consists of five identical branches, therefore the fan-out is equal to $n = 5$. Each load circuit contains two identical input branches consisting of transistors Q4 and Q5, so the fan-in of the load circuit is $m_L = 2$. The index "L" designates the load circuits in Fig. 2.34. The input voltage in the second part of the load circuit, transistor Q5, is equal to the voltage in the input of the first part, transistor Q4. All transistors and associated resistors in the circuit are identical. The values of the parameters of the circuit such as R_b, R_L, I_s, α_F, α_R are defined.

For simplicity, we exclude two of three inputs from consideration. For this purpose, we shorted out these inputs to ground. Another way is to short these inputs with the remaining working input. Also, we consider only one part of the load circuit. So, we obtain the simplified circuit presented in Fig. 2.35, which consists of only two transistors. Now we can analyze the circuit behavior.

In the initial state V_{in} is zero, therefore the transistor Q1 is in the cutoff mode and all the current flowing through the R_L supplies five load branches, that is, the transistor Q4 is in the saturation mode. In this state, $V_{in} = 0 = $ "0" and $V_o = V_{OH} = $ "1". As the input voltage increases, transistor Q1 starts to open and takes a part of the current. Now, the V_o decreases due to the voltage drop on the resistance R_L and transistor Q4 transits from the saturation mode to the linear mode. So, both transistors Q1 and Q4 are in the linear mode. The input voltage continues to grow and the output voltage V_o reaches

Table 2.3.   RTL circuit behavior.

Break points	Inverter (Transistor Q1)	Load (Transistor Q4)
1 $V_{BE,1} < V_{BE,\gamma}$	cutoff **BP 1**	saturation
2 *BP1* to *BP2*	linear **BP 2**	saturation
3 *BP2* to *BP3*	linear **BP 3**	linear
4 *BP3* to	linear **BP 4**	cutoff
5 $V_{BE,1} > V_{BE,sat}$	saturation	cutoff

the value closing transistor Q4. The additional increasing of V_{in} leads to the transition of transistor Q1 into the saturation mode. The described behavior may be approximately presented with the transfer function using the piecewise approach. We have two transistors, each of them has two break points as was mentioned above. Therefore, in the circuit's behavior, there are four break points. We present the circuit's behavior with these break points in Table 2.3.

Now we can calculate the coordinates of all four break points:

1. *BP1*

As shown in the first line of the table, transistor Q1 is in the cutoff mode and transistor Q4 is in the saturation mode up to the point that V_{BE1} is more than $V_{BE,\gamma}$. So, bearing in mind that $i_C = 0$ and using Kirchhoff's law, we can write the following equations:

$$V_{CC} = i_L R_L + \frac{i_L R_b}{n} + V_{BEL}^s \qquad (2.40)$$

$$V_o = V_{CC} - i_L R_L \qquad (2.41)$$

where V_{BEL}^s designates the saturation voltage of transistor Q4. Using the value of i_L from Eq. (2.40), we obtain for V_o:

$$V_o = \frac{V_{CC} R_b + n V_{BEL}^s R_L}{n R_L + R_b}. \qquad (2.42)$$

As we decided above, in the *BP1*, transistor Q1 transits from the cutoff mode to the linear mode and this point is defined by the beginning of the collector current flow through the transistor, $0.01 I_{max}$. Therefore, at this point $i_C = 0.01 V_{CC}/R_L$. Using the Ebers–Moll coupling Eq. (2.39), we can estimate the voltage on the

BE-junction of the transistor Q1:

$$V_{BE} = V_t ln \left(\frac{i_C}{\alpha_F I_{ES}} \right) = V_t ln \left(\frac{0.01 V_{CC}}{\alpha_F I_{ES} R_L} \right). \qquad (2.43)$$

Then, using the input Eq. (2.35) and Eq. (2.21) for the linear mode of operation for the transistor Q1, we obtain the following input voltage:

$$V_{in} = i_b R_b + V_{BE} = \frac{i_C R_b}{\beta_F} + V_t ln \left(\frac{0.01 V_{CC}}{\alpha_F I_{ES} R_L} \right). \qquad (2.44)$$

Thus, we calculated the coordinates of *BP1*: Eqs. (2.42) and (2.44).

2. *BP2*

Here the transistor Q1 is in the linear mode and the transistor Q4 is beginning to shift to the border of its linear region. So, to find the output voltage V_o, we need to calculate the total consumption of the load. A current flowing through Q4 is equal to:

$$i_{CL} = \frac{V_{CC} - V_{oL}^s}{m_L R_L} \qquad (2.45)$$

where $V_{OL}^s = V_{CE,sat} \approx 0.2\ V$ is the minimum possible voltage, $V_{CE,sat}$. Using the Ebers–Moll coupling Eq. (2.39) we can find the voltage drop on the BE-junction of transistor Q4:

$$V_{BEL} = V_t ln \left(\frac{i_{CL}}{\alpha_F I_{ES}} \right) = V_t ln \left[\frac{(V_{CC} - V_{oL}^s)}{\alpha_F I_{ES} m_L R_L} \right]. \qquad (2.46)$$

As transistor Q4 enters the linear mode of operation at the break point *BP2*, the collector current begins to become proportional to the base current:

$$i_{BL} = \frac{i_{CL}}{\beta_F} = \frac{V_{CC} - V_{oL}^s}{m_L R_L \beta_F}. \qquad (2.47)$$

In this way we can use the input Eq. (2.35) to estimate the output voltage V_o that is the input voltage to the load scheme:

$$V_o = i_{bL} R_b + V_{BEL} = \frac{(V_{CC} - V_{oL}^s) R_b}{m_L R_L \beta_F} + V_t ln \left[\frac{(V_{CC} - V_{oL}^s)}{\alpha_F I_{ES} m_L R_L} \right]. \qquad (2.48)$$

Now, using Eq. (2.48), we can calculate the collector current i_C:

$$i_C = \frac{V_{CC} - V_o}{R_L} - n i_{BEL} = \frac{V_{CC} - V_o}{R_L} - \frac{n(V_{CC} - V_{oL}^s)}{m_L R_L \beta_F}. \qquad (2.49)$$

Then, repeating steps (2.43) and (2.44) we will calculate the input voltage, V_{in}, associated with the break point *BP2*.

3. *BP3*

At this point, the transistor Q4 transits in the cutoff mode. Therefore, the collector current $i_{CL} \approx 0.01 V_{CC}/R_L$ and the base current $i_{bL} \approx 0$. In this state the output voltage V_o will be equal to the voltage drop of transistor Q4 at the BE-junction:

$$V_o = V_t ln \left(\frac{0.01 V_{CC}}{\alpha_F I_{ES} R_L} \right). \qquad (2.50)$$

As $i_L \approx 0$ and the collector current $i_C = (V_{CC} - V_o)/R_L$, we can repeat steps (2.43) and (2.44) to calculate the input voltage, V_{in}, associated with the break point *BP3*.

4. *BP4*

At this point, transistor Q1 enters the saturation mode. Therefore, $V_o = V_{CE,sat} = 0.2$ V. The collector current $i_C = (V_{CC} - V_{CE,sat})/R_L$. Repeating steps (2.43) and (2.44), we can calculate the input voltage, V_{in}, associated with the break point *BP4*.

We have calculated the coordinates (V_{in}, V_{out}) for all four break points characterizing the circuit in Fig. 2.34. Now, we can draw the transfer function calculated using the piecewise approach and estimate the noise margins.

2.4. The diode-transistor logic (DTL) family

As the title suggests, this family of logic circuits is based on both diodes and transistors. The following picture presents the typical DTL circuit.

The circuit represents a logic device with two inputs and one output. In terms of logic, this circuit consists of two parts: a logic

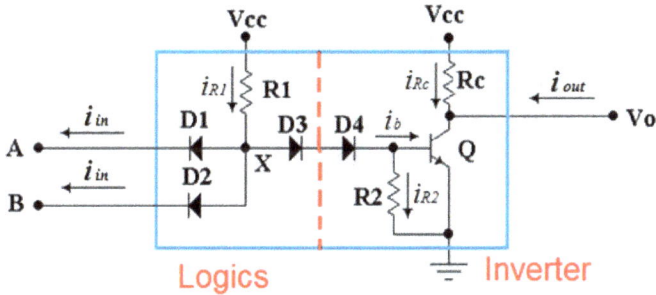

Fig. 2.36.   A typical DTL circuit.

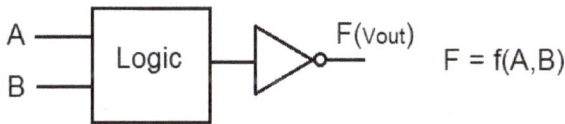

Fig. 2.37.   General view of a logic circuit.

input circuit and an inverter. It should be noted that most logic circuits can be represented thus regardless of technological bases and logic family type. In this way, the scheme presented in Fig. 2.36 may be presented in general view as shown in Fig. 2.37. In other words, each logic gate may be presented as a logic part and an inverter concluding the circuit.

The usual way to analyze the logic circuit consists of several steps:

1. Identify a logic function that the circuit performs;
2. Calculate all currents and voltages in the circuit for two different states of the output voltage;
3. Calculate the maximum fan-out value for this type of circuit;
4. Build the transfer function for the circuit on the condition that only one input is acting and estimate the noise margins.

Now, we will begin to analyze the circuit shown in Fig. 2.36 using the approach described. This circuit is logical, therefore the input signals can be in two states only: a logic "0" and logic "1". We will only use the intermediate values of input variables for building the transfer function. Usually, the initial data for such circuits are as follows: V_{CC}, $V_{D,\gamma}$, $V_{D,on}$, $V_{BE,\gamma}$, $V_{BE,on}$, $V_{CE,sat}$, β_F, and β_R. The simplest way to identify a logic function is to build a truth table. Two inputs define four possible states of the output voltage; however, the output function may take only two possible values: "1" and "0". So, we should analyze the circuit behavior in various combinations of inputs according to the following table:

N	A	B	V$_0$
1	0	0	
2	0	1	
3	1	0	
4	1	1	

It must be noted that consideration of the circuit in Fig. 2.36 shows that the input currents i_{in} and the output current i_{out} are directed in the right-to-left direction. As the digital circuits are typically definitely oriented on the charts: the inputs are on the left, the outputs are on the right, the power supply is at the top and the ground is at the bottom, so then the input currents in Fig. 2.36 are flowing out of the circuit and the output current flowing into the circuit. Also, as the output voltage is the collector-emitter voltage of the transistor, the minimum possible output voltage will be the saturation voltage of the transistor, or $V_{CE,sat}$. Evidently, in digital circuits, the input voltage is the output voltage of the previous logic gate of the same logic family. Therefore, the logical "0" $= V_{CE,sat}$ in the DTL family and the logical "1" $= V_{CC}$.

Let us suppose that $V_{CE,sat} \approx 0.2$ V, $V_{D,\gamma} = V_{BE,\gamma} = 0.5$ V, and $V_{D,on} = V_{BE,on} = 0.7$ V, and let us then consider the first line in our truth table. Here, A $=$ B $=$ "0" $= V_{CE,sat} \approx 0.2$ V. In this state, the voltage at point X of the circuit will be under the potential of

$$V_X = V_{in} + V_{D,on} = \text{"0"} + V_{D,on} = 0.2 + 0.7 = 0.9 \text{ V}. \qquad (2.51)$$

To start conducting, transistor Q requires a voltage equal to $V_{BE,\gamma} = 0.5$ V on the BE-junction. For this to occur, at point X, the voltage must be equal to

$$V_X = V_{BE,\gamma} + 2V_{D,on} = 0.5 + 0.7 + 0.7 = 1.9 \text{ V.} \qquad (2.52)$$

However, we only have 0.9 V (see Eq. (2.51)), therefore the transistor is in the cutoff mode. Thus, the output voltage will be equal to V_{CC} or "1". In this state, only input currents are flowing in the circuit and we can calculate these currents with Kirchhoff's equation:

$$V_{CC} = V_{in} + V_{D,on} + 2i_{in}\text{R}1 \Rightarrow i_{in} = (V_{CC} - V_{in} - V_{D,on})/2\text{R}1. \qquad (2.53)$$

If we consider the second and third lines in the truth table, we can see that they are symmetric, as the names for the inputs were designated randomly. Even if one of the inputs is equal to "0", all the arguments we have given above, Eqs. (2.51), (2.52) and (2.53), will still be valid. The only difference here is that the input current is double Eq. (2.53). The output voltage will be "1" once more.

Now, we can consider the case of A = B = "1". In this case both input diodes are cutoff and the current through R1 flows in the right direction, through the base of the transistor. The current value may be estimated from the following equation, Kirchhoff's equation:

$$V_{CC} = i_{R1} + V_{D,onD3} + V_{D,onD4} + V_{BE,sat}. \qquad (2.54)$$

Here, the transistor is in saturation mode and the output voltage $V_o = V_{CE,sat}$. If the supply voltage is not enough to bring the transistor into saturation, its value was chosen incorrectly. **In other words, the supply voltage should place the transistor in saturation mode.** The currents in this case will be as follows:

$$i_b = \frac{V_{CC} - 2V_{D,on} - V_{BE,sat}}{R1} - \frac{V_{BE,sat}}{R2}. \qquad (2.55)$$

$$i_C = \frac{V_{CC} - V_{CE,sat}}{R_c}. \qquad (2.56)$$

Now, we can fill the truth table. As shown in the truth table, our logic gate implements a function NAND. Since any logic function may

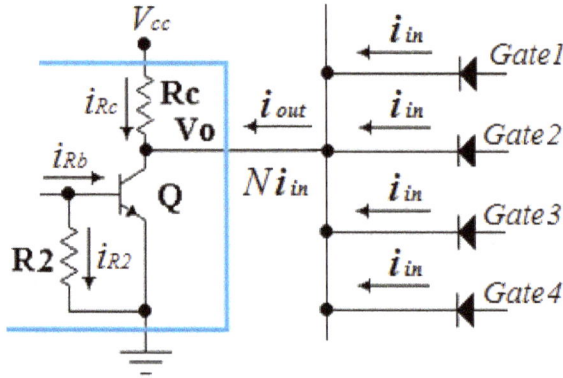

Fig. 2.38. Calculation of the maximum fan-out for DTL logic family.

be represented by the function NAND, the DTL circuit presented in Fig. 2.36 is generic.

N	A	B	V_0
1	0	0	1
2	0	1	1
3	1	0	1
4	1	1	0

We did two parts of the logic gate analysis: we found the logic function and we calculated all the currents in the circuit for two of its states. Our following task is to calculate the maximum fan-out for the circuit when connected to circuits from the same logic family. According to the definition in Chapter 1, the fan-out maximum is the number of inverters from the same logic family which can load the logic gate without changing the logic. In other words, this is an integer number of inverters which will not remove the transistor from the saturation state. One additional loading inverter will take the transistor to the linear mode. Figure 2.38 illustrates this calculation for N logic gates connected with the circuit output in parallel.

It is known that the output voltage in the digital logic gate may take only two different values: "1" and "0". If V_o = "1", all load diodes will be in the cutoff mode and their number will not matter. If the output voltage is "0", all load currents (input currents of the load inverters) will enter the transistor. In response to too much load, the transistors respond by increasing their resistance and moving from saturation to linear mode. Thus, the maximum number of load gates will meet at the transition point from saturation to linear mode, providing Eq. (2.21), $i_c = \beta i_b$. This equation may be rewritten for our calculation as follows:

$$\beta i_b = i_c = N i_{in} + i_{Rc}. \tag{2.57}$$

Therefore, the fan-out maximum for the DTL NAND circuit shown in Fig. 2.36 will be as follows:

$$N_{max} = Int\left\{\frac{\beta i_b - i_{Rc}}{i_{in}}\right\}. \tag{2.58}$$

The result in this equation should be rounded down.

The last parts of the analysis are the calculation of the break points, the drawing of the transfer function and the estimation of the noise margins. We have only one transistor in the DTL inverter as shown in Fig. 2.36. So, it only has two break points to fully define the transistor's behavior.

Point A designates the transition of the transistor from the cutoff to the linear mode of operation. At this point, the internal resistance of the BE-junction of the transistor decreases and the transistor begins to conduct. This point has been previously identified as $V_{BE} = V_{BE,\gamma} = 0.5$ V for silicon transistors. To obtain this value of V_{BE}, the voltage in the point V_X should be equal to $V_X = 2V_{D,on} + V_{BE,\gamma} = 2*0.7 + 0.5 = 1.9$ V. In our circuit: $V_{in} = V_A = V_B = V_X - V_{D,on} = 1.9 - 0.7 = 1.2$ V. Since this voltage marks the beginning of the transition process in the inverter, this voltage will be called V_{IL}. Figure 2.39 illustrates our calculations.

The transition process will continue until the transistor goes into saturation mode. This point is designated as B on Fig. 2.39. After that, our inverter will be in the state "0". As the transistor enters into saturation with the voltage $V_{BE} = 0.7$ V, the input voltage relating to point B will be equal to $V_{IH} = 1.4$ V. The argumentation is similar to that underlying the calculation of point A. Now we

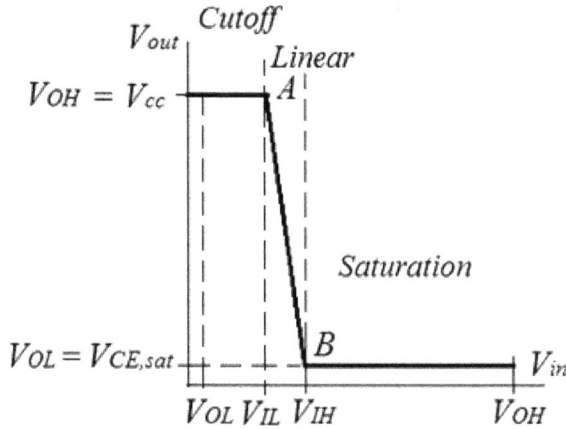

Fig. 2.39.   The transfer function of the DTL inverter.

can find the noise margins as shown in Fig. 2.33, on condition that $V_{CC} = 5$ V:

$$NM_L = V_{IL} - V_{OL} = 1.2 - 0.2 = 1 \text{ V}. \qquad (2.59)$$

$$NM_H = V_{OH} - V_{IH} = 5 - 1.4 = 3.6 \text{ V}. \qquad (2.60)$$

2.5.   The transistor-transistor logic (TTL) family

2.5.1.   *The basic TTL logic circuit*

The DTL logic has several deficiencies, such as high dimensions, high power consumption, non-symmetric transfer functions, etc. However, this logic family was a prototype for TTL, the first logic family created by applying integrated circuits built using planar technology. In 1965, "Texas Instruments" presented their logic series 7400 which became the industrial standard of logic circuits. The main basic logic gate in this family was the two-input NAND gate. However, the main difference between this gate and the others was the replacement of an input diode assembly on the novel semiconductor device — the multi-emitter transistor, as shown in Fig. 2.40.

This transistor may be considered approximate to the diode assembly, thus the simplest TTL logic gate looks like the DTL

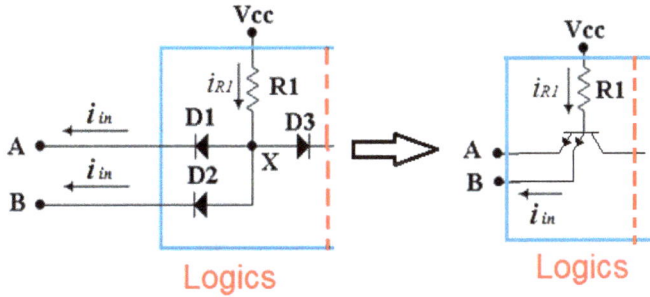

Fig. 2.40. A basic element of TTL logic — the multi-emitter transistor.

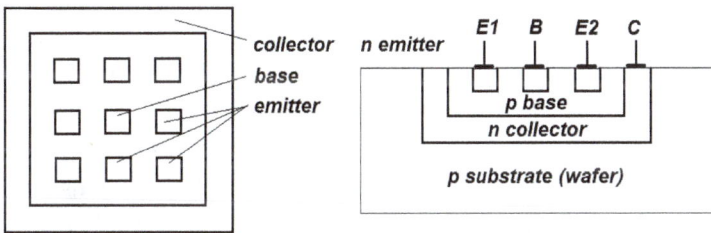

Fig. 2.41. A schematic presentation of the multi-emitter transistor.

inverter (see Fig. 2.36). The input transistor may be schematically presented as shown in Fig. 2.41. Here, two projections of the novel transistor are shown: the top view and cross-section of the transistor built by planar technology. All the distances between the elements on the picture are defined by the lithography step, λ. As shown, the fan-in of the input part of the logical gate is equal to 8. It should be noted that none of the non-connected inputs conducts any current. Therefore, the non-connected inputs behave like inputs connected to high voltage. They can pick up electromagnetic fields from air, like antennas, and cause harmful interference to normal operations. Therefore, in digital circuits, the free inputs should always be connected to the power supply via a separate resistor.

After the replacement of the input part, our logic gate may be seen in Fig. 2.42. Here, we see the simplest TTL logic gate implementing the function NAND. The analysis of this circuit is the same as in the case of the DTL inverter; however, there is one difference. As shown

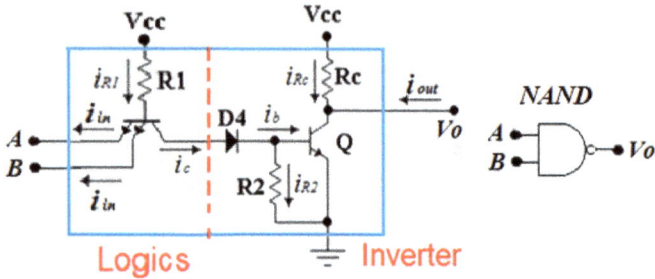

Fig. 2.42. The simplest TTL NAND gate.

in Fig. 2.42, the base voltage is always greater than the collector voltage. The input voltage, $V_{in} = V_A = V_B$, may be less or more than the base voltage. If the input voltage is less than the base voltage, the transistor will be in the saturation state. The input voltage will be equal to "1" (more than the base voltage), and the transistor will switch to the reverse linear mode (see Table 2.2). In this case, the calculation of the circuit currents will be different than in the case of the DTL inverter.

Let us assume that $V_{in} = V_A = V_B = V_{CC}$, so that the input transistor will be in the reverse active mode. Due to Kirchhoff's law, we can write the following equation:

$$i_c = i_{R1} + 2\beta_R i_{R1} = i_{R1}(2\beta_R + 1). \tag{2.61}$$

Therefore, the base current in the transistor Q will be different for TTL and DTL logic circuits. All the following analysis is the same in these circuits.

2.5.2. *The "totem-pole" TTL logic circuit*

The most common pattern in the TTL logic is called the "totem-pole" scheme. It is presented in Fig. 2.43. This scheme got its name due to the fact that all devices in the output circuit are on the same vertical line, as in the Indian totem. Two diodes, D_2 and D_3, represent protection against the voltage fluctuations which may occur on the ground bus.

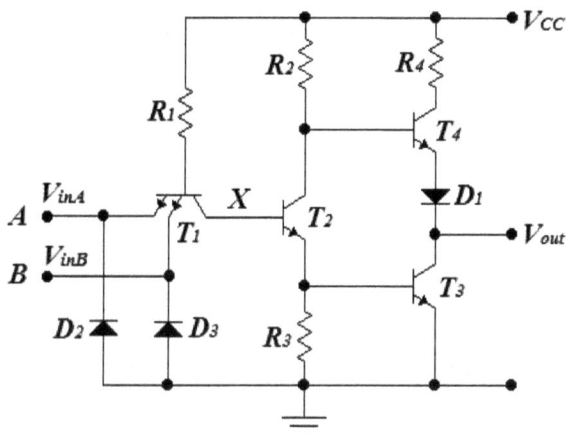

Fig. 2.43. The basic TTL circuit.

We will provide a full, step-by-step analysis of this logic gate in the order described above. Firstly, we should evaluate this circuit qualitatively to find the logic function performed by the scheme. To achieve this goal, we will consider the extreme states of the input terminals. If both inputs are in the low state, $V_{in} = V_A = V_B = $ "0", the input transistor will be in the saturation state and the voltage at point X will be $V_X = V_{in} + V_{CE,1}$. As the output voltage is equal to $V_{CE,3}$ (see Fig. 2.43), V_{out} may take two values only: "0" or "1". Therefore, V_{in} may also take only these values and the output voltage "0" $= V_{CE,sat}$. Thus, the voltage $V_X = 2V_{CE,sat}$ and $V_X < V_{BE,\gamma}$ of the transistor T_2. Therefore, transistor T_2 is in the cutoff state and consequently, transistor T_4 is in the saturation state and transistor T_3 is in the cutoff state. So, the output voltage is in the level of "1". It is interesting that all these arguments are true also in the case that only one of the input terminals is in the low state. The second input, which is in the "1" state, does not affect the input transistor behavior.

It is of interest to know what the maximum output voltage V_{OH} is which corresponds to the logical "1" in this state. As transistor T_3 is in the cutoff state and all loading circuits are connected with our circuit using the emitter's terminals, the current practically does not flow through the diode D_1 and the output terminal. Therefore, in condition that $V_{CC} = 5$ V, the output voltage may be approximately calculated in the following way:

$$V_{out\text{"1"}} = V_{OH} = V_{CC} - V_{BE4,on} - V_{D1,on} = 5 - 0.7 - 07 = 3.6 \text{ V}.$$

If both the input terminals are in the high state, "1", the input transistor switches to the reverse linear mode and all currents from the inputs and the BC junction of transistor T_1 enter into the base of transistor T_2. Transistor T_2 switches to the saturation mode and shorts the bases of transistors T_3 and T_4. Transistor T_3 enters the saturation state that is $V_{out} = V_{CE,sat3} = 0.2$ V and transistor T_4 enters the cutoff, due to a low voltage on its base: $V_{B,4} = V_{BE,on3} + V_{CE,sat2} = 0.7 + 0.2 = 0.9$ V which is lower than the voltage required for the activation of transistor T4:

$$V_{B,4} = V_{out} + V_{D1,on} + V_{BE,\gamma4} = 0.2 + 0.7 + 0.5 = 1.4 \text{ V}.$$

Therefore, the logic gate presented in Fig. 2.43 represents a gate NAND and its truth table is as follows:

N	A	B	V_0
1	0	0	1
2	0	1	1
3	1	0	1
4	1	1	0

The second step of the analysis is the calculation of all the voltage drops and currents in the circuit for two different states of the gate and the estimation of the maximum fan-out. This analysis is usually a fairly routine calculation and here we will consider the difference in the estimation of the maximum fan-out. We consider later on a concrete example upon which detailed analysis has been performed. If we compare the two circuits shown in Figs. 2.38 and 2.43, we can see that calculation of the maximum fan-out makes sense only in the case of the low state of the circuit, $V_{out} = V_{CE,sat} = 0.2$ V. In both circuits, the output transistor consumes the input currents of the load inverters, however in the TTL "totem-pole" the closed transistor T_4

Table 2.4.   Totem-pole circuit behavior.

Breakpoints	T1	T2	T3	T4
1 $V_{BE2} \leq V_{BE2,\gamma}$	Saturation	Cutoff	Cutoff	Saturation
2 BP$_A$ to BP$_B$	Saturation	Active	Cutoff	Saturation
3 BP$_B$ to BP$_C$	Saturation	Active	Active	Active
4 $V_{BE2} \geq V_{BE2,on}$	Reverse active	Saturation	Saturation	Cutoff

(with **BP$_A$** between rows 1–2 at T2, **BP$_B$** between rows 2–3 at T3, **BP$_C$** between rows 3–4 at T4)

prevents the current from flowing through the resistor R_4. Therefore, the equation for the calculation of the maximum fan-out, Eq. (2.58), transforms as follows:

$$N_{max} = Int \left\{ \frac{\beta i_b}{i_{in}} \right\}. \tag{2.62}$$

The result in this equation should be rounded down.

Now, we can calculate the break point coordinates and prepare the transfer function. As the circuit in Fig. 2.43 consists of four transistors, we should take into account all the transistors. However, the following reasoning enables us to prepare a simplified transfer function using three break points only. Table 2.4 represents the transistor behavior during the change in the circuit from "1" to "0".

The first extreme state of the circuit, "1", that is defined by low input voltage, continues until the input voltage does not exceed the value required to change the mode of transistor T_2 from cutoff mode to active mode: $V_{BE2} \leq V_{BE2,\gamma}$. This is achieved by the following equation:

$$V_{IL} = V_{in} + V_{CE1,sat} = V_X = V_{BE2} = 0.5 \ V.$$

Thus, the coordinates of the first break point, A, are:

$$(V_{inA}, V_{outA}) = (V_{IL}, V_{OH}) = (0.3, 3.6).$$

Now, after the break point A, the current through the transistor T_2 will increase when the input voltage is increased. So, the potential of the base of transistor T_4 will decrease and the potential of the base of transistor T_3 will increase. The break point B is characterized by the transition of the transistor T_3 from cutoff to active mode. To enable this transition, the input voltage should be provided by $V_{BE3} = V_{BE3,\gamma}$:

$$V_{inB} = V_X - V_{CE1,sat} = V_{BE2,on} + V_{BE3,\gamma} - V_{CE1,sat} = 0.7$$

$$+ 0.5 - 0.2 = 1 \text{ V.}$$

Now, let us find the value V_{outB}. In this interval, A-B, a current flowing through resistors R_2 and R_3 will be practically equal, in other words, if we neglect the current through the base of transistor T_2, one can say that the collector current and the emitter current of the transistor T_2 are approximately equal, or $i_{c2} \approx i_{e2}$.

$$i_{e2} = \frac{V_{e2}}{R_3} = \frac{V_{B2} - V_{BE2,on}}{R_3} = \frac{V_{in} + V_{CE1,sat} - V_{BE2,on}}{R_3}. \qquad (2.63)$$

$$i_{c2} = \frac{V_{cc} - V_{B4}}{R_2} = \frac{V_{cc} - V_{out} - V_{D1,on} - V_{BE4,on}}{R_2}$$

$$= \frac{V_{in} + V_{CE1,sat} - V_{BE2,on}}{R_3} = i_{e2}. \qquad (2.64)$$

The differential of Eq. (2.63) gives us a transfer function in the interval A-B:

$$\frac{dV_{in}}{R_3} = -\frac{dV_{out}}{R_2} \Rightarrow \frac{dV_{out}}{dV_{in}} = -\frac{R_2}{R_3} = const. \qquad (2.65)$$

Therefore, the transfer function in this interval represents a direct line or a linear function. The exact value of the voltage V_{outB} may be calculated using Eq. (2.65) as follows:

$$\frac{V_{OH} - V_{outB}}{V_{IL} - V_{inB}} = -\frac{R_2}{R_3}. \qquad (2.66)$$

The break point C may be defined by the transition of transistor T3 to the saturation mode. At this point, the voltage V_{out} reaches its minimum value $V_{CE3,sat}$ and the circuit enters the logical "0" state. The input voltage relating to this state is calculated as follows:

$$V_{in} = V_{IH} = V_{BE3,on} + V_{BE2,on} - V_{CE1,sat} = 1.2 \text{ V.}$$

Therefore, the coordinates of the break point C will be $(V_{IH}, V_{OL}) = (1.2, 0.2)$. The transfer function built using the above calculation is shown in Fig. 2.44.

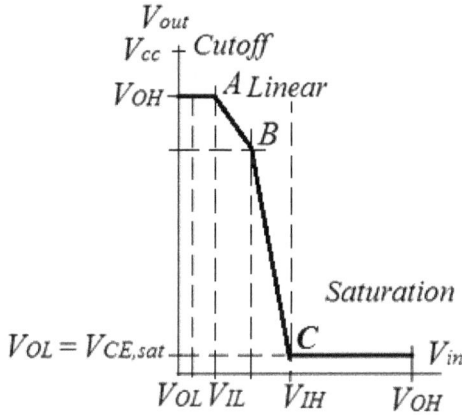

Fig. 2.44.   The transfer function of the "totem-pole" circuit.

2.5.3.   *Further improvements of TTL logic circuits*

One of the main problems of the TTL logic family is how slowly it switches. This problem is caused by the accumulation of a large electrical charge in the base region of the transistor which is in the saturation state. This charge requires a long time to discharge the base capacitance while the transistor is switching. A good solution to this problem is not to let the transistor go into a deep saturation state. Technically, this problem can be solved by the parallel connection of the Schottky diode to the base-collector junction of the transistor as shown in Fig. 2.45. Here, the metal, when connected to the base (a p-type semiconductor), forms an Ohmic contact with the base. At the same time this metal forms a Schottky diode with the collector, which is an n-type semiconductor, see Fig. 2.45(c). Figure 2.45(b) represents a transistor with the Schottky diode and Fig. 2.45(a) designates such a transistor.

A transistor with a Schottky diode works as follows: when the voltage on the base begins exceed over the collector voltage, the Schottky diode transits into the conductive state and shorts the base-collector junction. Thus, the transistor cannot enter into the deep saturation state. A typical TTL NAND gate using the Schottky diodes to speed up its performance is shown in Fig. 2.46.

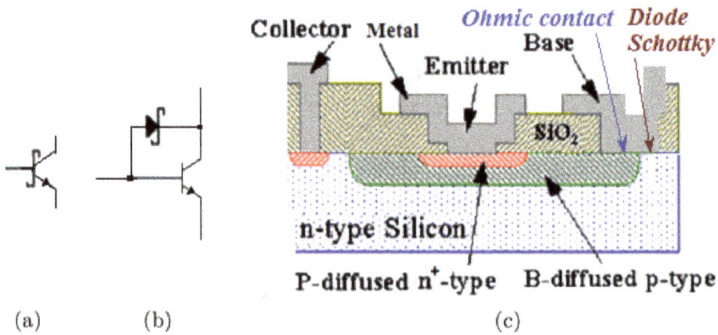

Fig. 2.45.   The application of the Schottky diode in the BJT transistor.

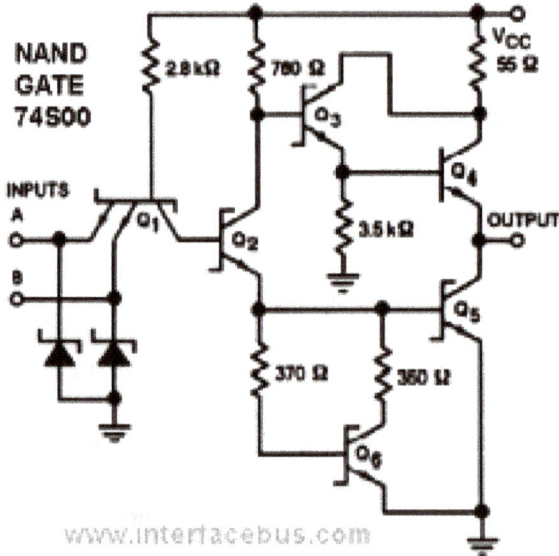

Fig. 2.46.   A typical TTL NAND gate using the Schottky diodes.

2.5.4.   *Three-state TTL logic circuits*

All logic gates, in the steady state, can be in one of two states: a logical zero, "0", or logical one, "1". To connect two or more outputs of logic gates, we can short an output which is in the high state with an output which is in the low state. In this case the circuit will not

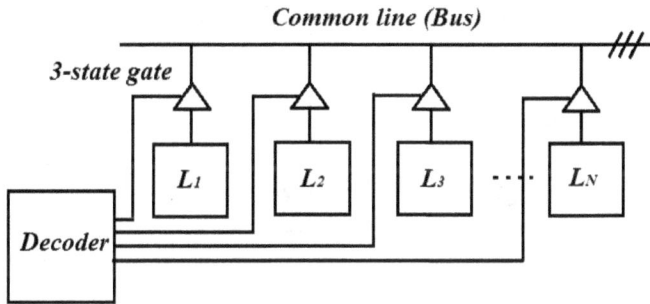

Fig. 2.47.   A circuit with three-state gates.

operate correctly. So, if several outputs of different logic circuits are connected together, we must ensure the operation of only one circuit and shut off all the other circuits. An example of such a system is the connection of various parts of a computer to its internal common bus. To ensure the smooth operation of a computer, all the components of the computer such as a memory, various registers, and processors are connected to the internal common bus through specific logic gates called three-state circuits or three-state gates. Figure 2.47 illustrates one of these systems.

In this picture, several logic gates, L_i, are connected on the same wire through specific logic gates which have three states: logic zero, logic one, and a third specific state when the output terminal is not connected to either "0" or "1" and is in the high impedance state. We can choose to connect one of the gates to the common line using the **Decoder**. There are various logic gates enabling three states: two logical states and one with high impedance. One of these logic circuits, the NAND gate with additional input, is shown in Fig. 2.48.

This circuit represents the usual TTL NAND logic gate (see Fig. 2.43) with one additional input, E, and two additional diodes, D_4 and D_5. Let us consider the operation of the gate presented in Fig. 2.48. If the input voltage supplied by the terminal E is in the high state, "1", the diodes D_4 and D_5 will be in cutoff mode. Also, the input emitter of the transistor T_1, which is connected to the terminal E, will be in the high state. Now, all the operations of

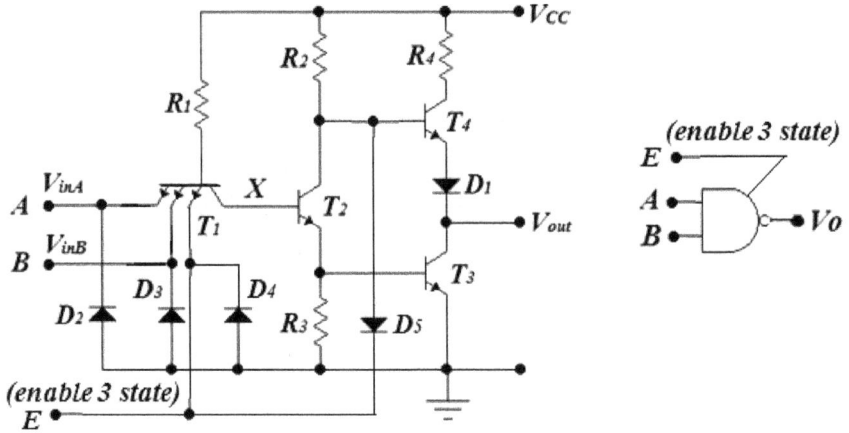

Fig. 2.48.   A TTL logic gate with three states.

the circuit are defined by the states of the input logic terminals: A and B. So, the terminal $E = $ "1" does not affect the operation of the circuit.

If $E = $ "0", the transistor T_1 is in the saturation mode. Therefore, as was mentioned for the circuit in Fig. 2.43, transistors T_2 and T_3 are in the cutoff mode and the terminal V_{out} is not connected to the low state (the transistor T_3 does not conduct any current). On the other hand, the diode D_5 will be in the active mode and the potential of the base of the transistor T_4 will be no more than 0.9 V ($V_{B4} = V_{in\,"0"} + V_{D5,on}$). This voltage is not enough to run the transistor T_4. Thus, the transistor T_4 is in cutoff mode and its collector is not connected to the supply. Therefore, the output terminal V_{out} of the circuit is separated from the "low state" and from the "high state" and it is in the high impedance state (the third state).

2.6.   The emitter-coupled logic family (ECL)

If we take two identical transistors and connect their emitters together as shown in Fig. 2.49, we obtain a circuit which can be applied in digital electronic systems.

As shown in Fig. 2.49, the currents flow through both transistors and we can write down the values of these currents using the

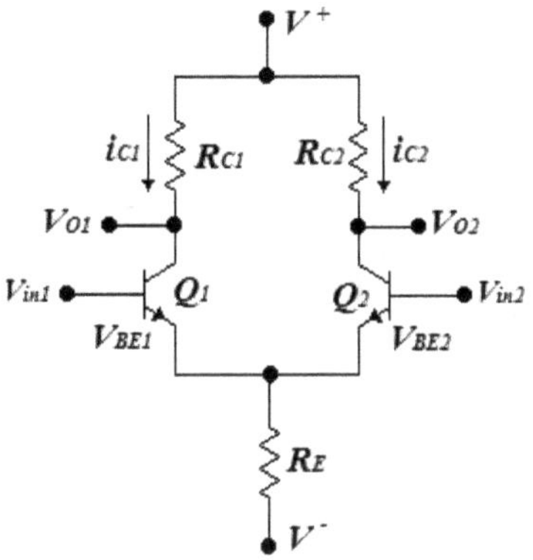

Fig. 2.49. Principal electrical scheme of the switch of currents.

approximated Shockley equation, as follows:

$$i_{C1} = I_s e^{\frac{V_{BE1}}{V_t}}. \tag{2.67}$$

$$i_{C2} = I_s e^{\frac{V_{BE2}}{V_t}}. \tag{2.68}$$

If we divide Eq. (2.67) by Eq. (2.68), we will obtain the following equation:

$$\frac{i_{C1}}{i_{C2}} = e^{\frac{V_{BE1}-V_{B2}}{V_t}}. \tag{2.69}$$

If we assume that $V_{BE1} - V_{BE2} = 0.12$ V, we obtain $i_{C1} = 101 \cdot i_{C2}$ at room temperature when $V_t = 0.026$ V, i.e. the current through the second transistor is more than 100 times less than the current through the first. Thus, the circuit shown in Fig. 2.49 represents a switch of currents. The current will flow through the transistor with the high input voltage and the second transistor will not conduct a current. The main requirement is to provide a difference between V_{in1} and V_{in2} which is more than 120 mV. Let us now modify the circuit in Fig. 2.49. We will fix the input voltage of the second transistor on the constant level $V_{in2} = V_R$, called the **reference voltage**, and put it in the approximate middle between two input voltages adopted for logic zero and logic one. The modified circuit is presented in Fig. 2.50.

Assume that the reference voltage in our modified circuit is -1.3 V, that the input voltage may take two extreme levels:

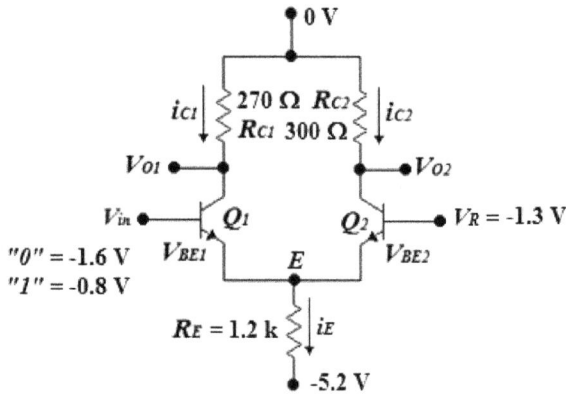

Fig. 2.50. The basic ECL inverter-buffer principal circuit.

"0" $= -1.6$ V and "1" $= -0.8$ V, that the base-emitter voltage $V_{BE} = 0.8$ V, and that the transistors have $V_{CE,sat} = 0.2$ V in saturation mode. Assume also that the input voltage is low, $V_{in} =$ "0" $= -1.6$ V. In this case, the difference between the base voltages $V_{in} - V_R = -1.6 - (-1.3) = -0.3$ V and $i_{C1}/i_{C2} \approx 1 \cdot 10^{-5}$. So, the current practically does not flow through the transistor Q_1. Instead, it flows through the transistor Q_2. We have the following states of the transistors: Q_1 is closed ("off") and Q_2 is open ("on"). The output voltage V_{O1} will be equal to 0 V since no voltage falls on the resistor R_{C1}. Now, one can calculate the output voltage V_{O2}, which is low due to the current flow through the resistor R_{C2}. The voltage at point E (see Fig. 2.50) will be $V_E = V_R - V_{BE,on} = -1.3 - 0.8 = -2.1$ V. Then the current $i_E = (V_E - V^-)/R_E = (-2.1 + 5.2)/1.2 = 2.6$ mA. If we neglect the base current, the emitter current will be equal to the collector current $i_{C2} = i_E = 2.6$ mA. Using this current, we calculate the output voltage $V_{O2} = 0 - i_{C2} \cdot R_{C2} = -2.6 \cdot 10^{-3} \cdot 300 = -0.8$ V. The voltage V_{CE} for the transistor Q_2 will be $V_{CE2} = V_{O2} - V_E = -0.8 - (-2.1) = 1.3$ V > 0.2V $= V_{CE,sat}$ which means that transistor Q_2 is in the active mode. The voltage V_{CE1} of the transistor Q_1 is equal to $V_{CE1} = V_{O1} - V_E = 0 - (-2.1) = 2.1$ V $> 0.2 = V_{CE,sat}$ which means that the transistor Q_1 is also in the active mode.

Table 2.5.   Logic signals in the ECL circuit.

V_{in}	V_{O2}	V_{O1}
"0" (-1.6 V)	"0" (-0.8 V)	"1" (0 V)
"1" (-0.8 V)	"1" (0 V)	"0" (-0.8 V)

Now, assume that the input voltage is high, $V_{in} =$ "1" $= -0.8$ V. In this case, the difference between base voltages $V_{in} - V_R = -0.8 - (-1.3) = 0.5$ V and $i_{C1}/i_{C2} \approx 2 \cdot 10^8$. So, the current practically does not flow through the transistor Q_2; instead, it flows through the transistor Q_1. We have the following states of the transistors: Q_2 is closed ("off") and Q_1 is open ("on"). The output voltage V_{O2} will be equal to 0 V since no voltage falls on the resistor R_{C2}. Now, we can calculate the output voltage V_{O1} which is low due to current flow through the resistor R_{C1}. The voltage at point E (see Fig. 2.50) will be $V_E = V_{in} - V_{BE,on} = -0.8 - 0.8 = -1.6$ V. Then the current $i_E = (V_E - V^-)/R_E = (-1.6 + 5.2)/1.2 = 3$ mA. If we neglect the base current, the emitter current will be equal to the collector current $i_{C1} = i_E = 3$ mA. Using this current, we calculate the output voltage $V_{O1} = 0 - i_{C1} \cdot R_{C1} = -3 \cdot 10^{-3} \cdot 270 = -0.8$ V. The voltage V_{CE} for the transistor Q_1 will be $V_{CE1} = V_{O1} - V_E = -0.8 - (-1.6) = 0.8$ V > 0.2 V $= V_{CE,sat}$ which means that transistor Q_1 is in the active mode. The voltage V_{CE2} of the transistor Q_2 is equal to $V_{CE2} = 0 - V_E = 0 - (-1.6) = 1.6$ V $> 0.2 = V_{CE,sat}$ that also means that the transistor Q_2 is in the active mode.

To summarize:

- Both transistors are always in the active mode and cannot enter into the cutoff or saturation modes, thus time to recharge the base regions in the transistors is not required in this circuit. Because of this, the ECL family is the fastest of all logic families;
- The circuit ECL (see Fig. 2.50) represents the digital switch of the currents and has two output voltages;
- Both outputs are always in opposite states, so we can call this circuit the inverter/buffer;
- The output voltages "0" and "1" are not compatible with the input voltages as shown in Table 2.5.

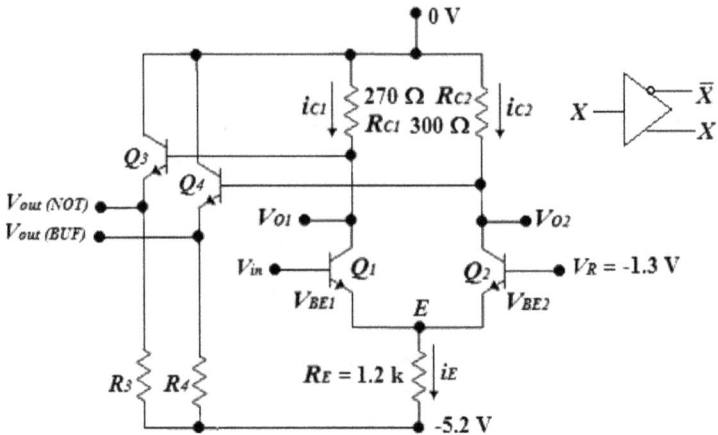

Fig. 2.51. The basic ECL circuit with compatible input-output terminals.

To solve the problem of the compatibility of input and output voltages, the circuit shown in Fig. 2.50 must be modified as shown in Fig. 2.51. The schematic designation of the circuit performing two functions, inverter and buffer, at the same time, is shown in this figure also.

In order to be able to design complex digital systems based on ECL technology, it is necessary that the level of the input and output voltages is the same. That is, we need to decrease the output voltage presented in Table 2.5. To achieve this goal, the output voltage V_{O1} and V_{O2} are connected with the emitter follower schemes based on transistors Q_3 and Q_4. These transistors are in the active mode and they are connected with the inverter/buffer through resistors R_3 and R_4 with resistance of about 1.5 kΩ. In this way, the high output voltage decreases to the value of the $V_{BE} = 0.8$ V and the low output voltage decreases to -1.6 V. Thus, we should obtain output voltages at the levels of "1" $= -0.8$ V and "0" $= -1.6$ V that are fully compatible with the input voltages.

To calculate the maximum fan-out for the typical ECL circuit, we must define what criterion will designate the maximum fan-out. Evidently, we can connect an infinite number of load circuits to an output terminal which is in the low state, as in this state the load transistors will be in the cutoff mode. Therefore, the maximum

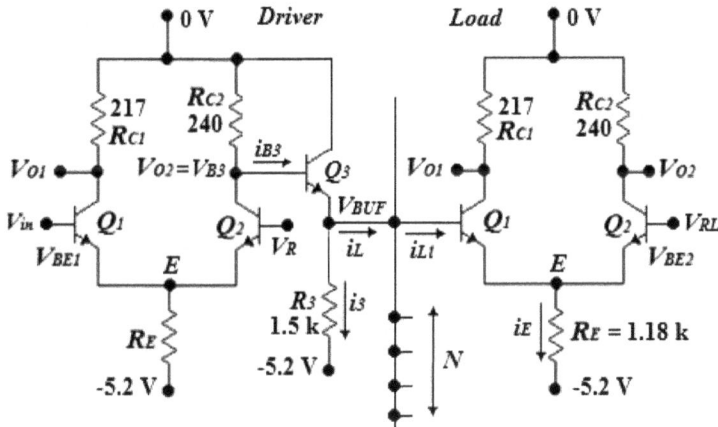

Fig. 2.52. A typical ECL inverter/buffer loaded with the same circuits.

fan-out calculation is meaningful only when the output terminal is in the high state. Each loading circuit will take a part of the output circuit and the output voltage will decrease until it is lower than the usual "1" voltage. We can set this value and calculate the maximum number of loading circuits so that the value of the output voltage does not drop below a predetermined value. Now, to calculate the maximum fan-out, let us consider Example 5.

Example 5.

Figure 2.52 represents a typical ECL inverter/buffer loaded with N of the same circuits.

Index "L" designates the parameters related to the load circuit. It is known that all transistors are the same with the gain $\beta = 50$ and $V_{BE,on} = 0.7$ V. Also, let us suppose that under loading by N load circuits, the value of logic one cannot decrease lower than 50 mV, therefore the output "buffer" is in the decreased high state $V_{BUF} = -0.75$ V instead of the "normal" "1" $= -0.7$ V. Here N is the maximal fan-out.

Solution 5.

As all transistors in the scheme are in the active mode, the base current of the transistor Q_1 in each loading circuit i_{L1} will be equal to its emitter current i_E/β. We can calculate the emitter current i_E from the following equation, known as Kirchhoff's equation:

$$i_E R_E = V_{BUF} - V_{BE,on} - V^- \text{ or } i_E = (-0.75 - 0.7 + 5.2)/1.18$$

$$= 3.18 \text{ mA}.$$

So, the base current $i_{L1} = 3.18/50 = 62.4$ μA and the total load current $i_L = i_{L1}N$.

A current i_3 flowing through the resistor R_3 is equal to $i_3 = (V_{BUF} - V^-)/R_3$ or

$$i_3 = (-0.75 + 5.2)/1.5 = 2.97 \text{ mA}.$$

Then we can calculate the base current i_{B3}:

$$i_{B3} = \frac{i_3 + i_L}{1 + \beta} = \frac{0 - V_{B3}}{R_{C2}} = \frac{0 - (V_{BUF} + B_{BE,on})}{R_{C2}}$$

or

$$\frac{2.97 + N \times 0.0624}{51} = \frac{-(-0.75 + 0.7)}{0.24}.$$

From these equations, we obtain the maximum fan-out $N = 122$.

Now we can build the transfer function for the ECL inverter/buffer. Once more, we should define when the transistor which is closed will begin to conduct the current (see Fig. 2.49). Let us agree that the transistor begins to transfer a current when the current reaches 1% of the current flowing through the opposite transistor. In other words, the current $i_{C1} = 99 \cdot i_{C2}$. This leads to the fact that the difference between V_{BE1} and V_{BE2} will be 0.115 V (see Eq. (2.69)). Using this value, we can find the values of V_{IL} and V_{IH} and build the transfer function for the ECL inverter-buffer. Figure 2.53 represents the transfer function for the ECL inverter-buffer circuit. The calculation of border values V_{IL} and V_{IH} is shown here:

$$V_{IL} = V_R - 0.115 = -1.3 - 0.115 = -1.415 \text{ V}.$$

$$V_{IH} = V_R + 0.115 = -1.3 + 0.115 = -1.185 \text{ V}.$$

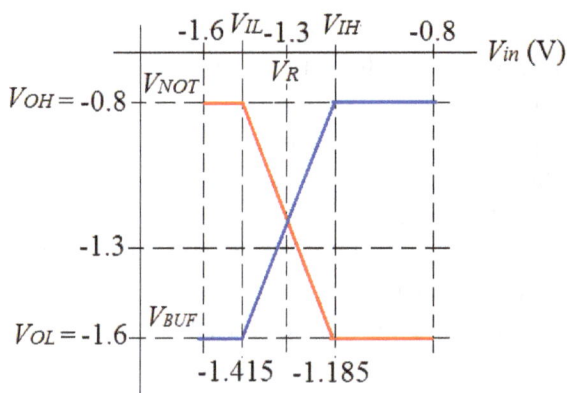

Fig. 2.53. Approximate transfer function of the ECL inverter/buffer.

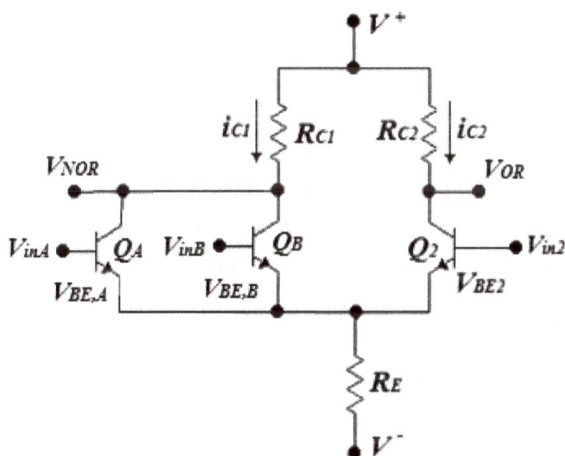

Fig. 2.54. A basic ECL logic gate OR/NOR.

The noise margins for these circuits may be calculated using Eqs. (2.70) and (2.71):

$$NM_L = V_{IL} - V_{OL} = -1.415 - (-1.6) = 0.155 \text{ V}. \qquad (2.70)$$

$$NM_H = V_{OH} - V_{IH} = 0.8 - (-1.185) = 0.385 \text{ V}. \qquad (2.71)$$

A basic logic gate in the ECL family is shown in Fig. 2.54. If we consider the circuit shown in Fig. 2.54, we can see that two inputs,

V_{inA} and V_{inB}, are symmetric and have been designated randomly. Therefore, no matter which of the transistors, Q_A or Q_B, we will analyze, the second will behave similarly.

As we have seen earlier, if one or both of the inputs are in the high state, the current will flow through the resistor R_{C1} and the transistor Q_2 will be closed. In this case, the output voltage V_{NOR} will be low and V_{OR} will be high. The transistor Q2 will conduct all current only if both inputs are low, $V_{inA} = V_{inB} = $ "0". These arguments can be illustrated using a truth table:

V_{inA}	V_{inB}	V_{NOR}	V_{OR}
0	0	1	0
0	1	0	1
1	0	0	1
1	1	0	1

Since any logic function can be built on the base of the logical gates NOR only, the ECL logic family is complete and self-sufficient. A typical ECL logic circuit is presented in Fig. 2.55. As shown, this circuit contains three symmetrical inputs and two matched

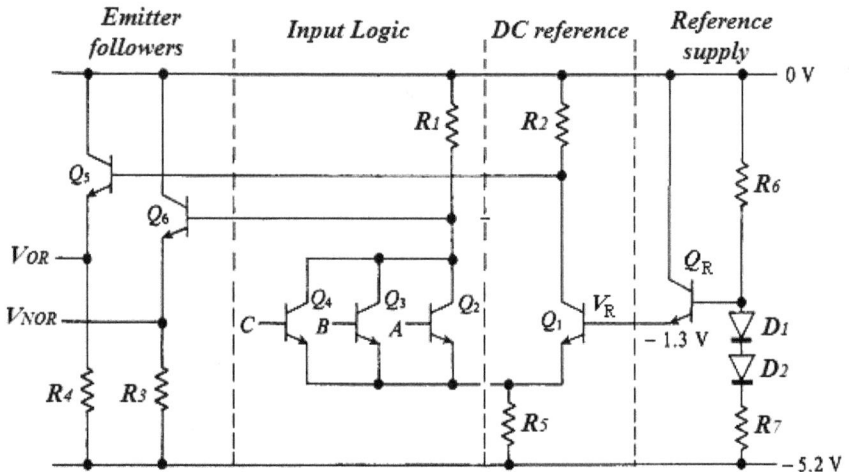

Fig. 2.55.   A typical ECL logic gate.

outputs, V_{NOR} and V_{OR}. In this circuit, there is one additional element designated as the reference supply. It comprises a transistor Q_R whose base is fed by the operating voltage of a voltage divider consisting of two resistors, R_6 and R_7, and two diodes, D_1 and D_2, which are both in the dynamic state (see Fig. 2.4). These silicon diodes act as negative feedback and provide the stable reference voltage V_R. They work as follows: if for any reason the emitter current of transistor Q_R is increased, this means that its base current will also increase, i.e., the voltage drop on the diodes and resistor R_7 increases also. Then, because of the non-linearity of the current-voltage characteristics of the diodes, their resistance will drop and the base voltage will also decrease and will return to the normal level.

2.7. The integrated injection logic family (I^2L)

One other group of digital circuits based on the use of bipolar junction transistors is the Integrated Injection Logic (IIL or I^2L) family. A simple inverter from this family is shown in Fig. 2.56.

In this circuit, the transistor T_1 is used as the stable current supply. Now, an input voltage V_A defines the status of the transistor T_2. If V_A is high, a current from the transistor T_1 enters the base of the transistor T_2 and puts it in the saturation state, provided that V_{out} is low. If the input voltage is low, the current from the transistor T_1 flows in the V_A direction and transistor T_2 enters cutoff mode, providing a high output voltage. Figure 2.57 represents a typical

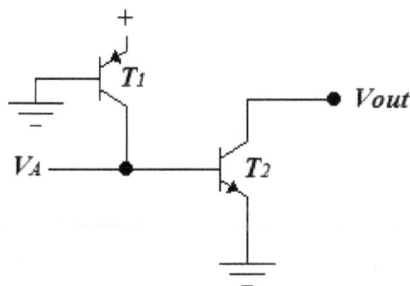

Fig. 2.56. I^2L inverter, a principle scheme.

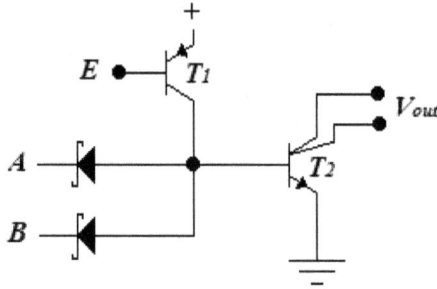

Fig. 2.57.   Typical NAND logic gate based on I^2L family.

NAND circuit equipped with the input E (enable) allowing the circuit to be transferred to the high impedance state.

In this circuit we have two logic inputs, A and B, providing the defined logic function in the case of the low state of the input E. If the voltage in the input E is high, the transistor T_1 will be in the cutoff state and will not provide supply for the base circuit of the transistor T_2. The behavior of the circuit shown in Fig. 2.57 may be illustrated using a truth table:

E	A	B	V_{out}
1	X	X	High impedance
0	0	0	1
0	0	1	1
0	1	0	1
0	1	1	0

2.8.   Problems

2.8.1 Assume that we have a PN-junction diode with the leakage current of 10^{-14} A at room temperature.

   a. Find an expression for the dynamic resistance of the diode and calculate its value for the following external voltages: 0.4, 0.5, 0.6, 0.7 V.

   b. Draw the chart of calculated dependence.

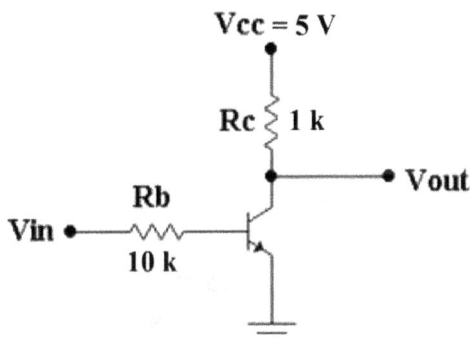

Fig. 2.58.   Typical RTL inverter.

2.8.2 The silicon PN diode consists of the N-layer with the donor concentration of $4 \cdot 10^{17}$ cm^{-3} and the P-layer with the acceptor concentration of $4 \cdot 10^{15}$ cm^{-3}. Calculate the capacity of the PN junction if we know that its interface area equals to 10^{-4} mm^2 and the relative dielectric constant of Si is equal to 11.7.

2.8.3 The diode from problem 2.8.2 is forward biased with a voltage of 0.6 V. Estimate the diffusion capacity of the diode at room temperature if we know that the mobility of charge carriers is equal to $\mu_n = \mu_p = 400$ cm^2/V·s, the lifetime of the minor charge carriers is 0.1 s and the length of both sides of the diode is 1 mm.

2.8.4 Calculate the average power dissipated by the RTL circuit presented in Fig. 2.58 in two different states: "1" and "0". Neglect power dissipation in the transistor.

2.8.5 The BJT transistor of NPN type, prepared from silicon, is at room temperature. The semiconductor layers of the transistor are doped in the following way: $N_{DE} = 5 \cdot 10^{17}$ cm^{-3}, $N_{AB} = 10^{16}$ cm^{-3}, $N_{DC} = 10^{15}$ cm^{-3}. It is known that the intrinsic concentration of free charged carriers is equal to $n_i = 1.5 \cdot 10^{10}$ cm^{-3} and the thickness of the base region is 0.8 μm.

   a. Calculate the concentration of minority carriers in the transistor's layers.

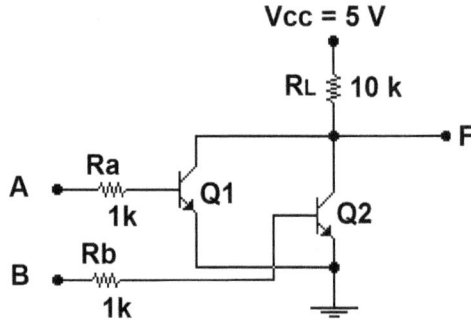

Fig. 2.59.   RTL logic circuit.

b. Calculate the built-in potential at both semiconductor junctions of the transistor and a suitable width for the depletion zones.

c. Assume that this transistor is connected to an electrical circuit by the scheme of the "common emitter" (see Fig. 2.58).

If we know that Rc = 1.5 kΩ, Vcc = 10 V, Rb = 0, Ic = 3 mA, and V_{in} = 0.8 V, calculate voltages V_{CB} and V_{CE}.

2.8.6 A logic circuit shown in Fig. 2.59 consists of two of the same transistors with the leakage current equal to 50 μA.

a. Identify the logic function implemented by the presented circuit.

b. Calculate the output voltage if both input voltages are low.

2.8.7 Calculate the noise margins for the circuit presented in Fig. 2.60 which works as a digital system if we know that $V_{D,on} = V_{BE,on} = 0.7$ V, $V_{BE,\gamma} = 0.5$ V, $V_{CE,sat} = 0.2$ V.

2.8.8 The circuit presented in Fig. 2.61 consists of elements with the following parameters: $V_{D,on} = V_{BE,on} = 0.7$ V, $V_{BE,\gamma} = 0.5$ V, $V_{CE,sat} = 0.2$ V, $\beta_F = 50$, $\beta_R = 0.1$.

a. Identify the logic function implemented by the presented circuit.

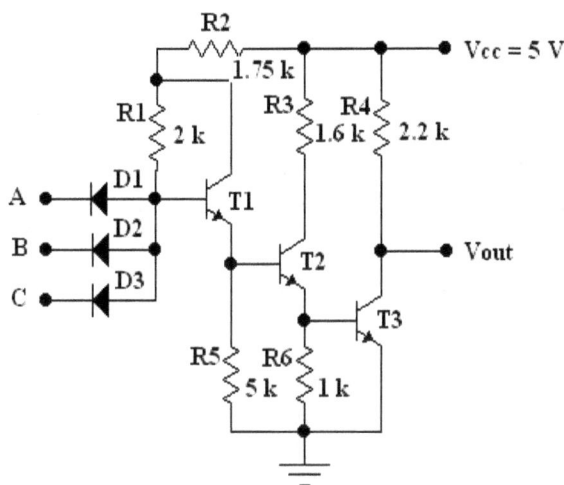

Fig. 2.60. DTL logic circuit.

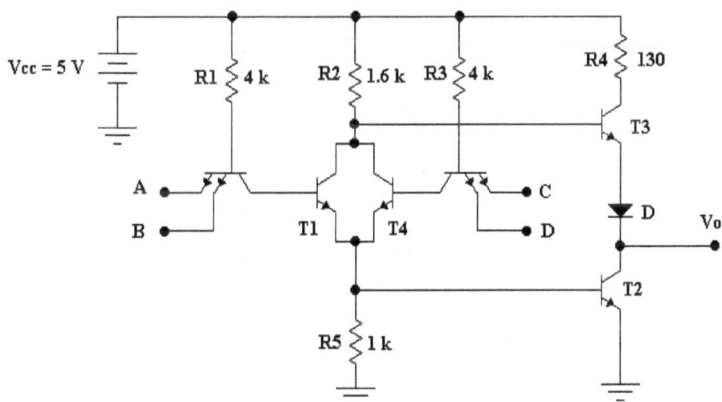

Fig. 2.61. TTL logic circuit.

b. Calculate the average power consumption of the circuit for two cases: all inputs at the low voltage "0" and all inputs at the high voltage "1".

c. Calculate the maximum fan-out for the presented circuit.

2.8.9 The circuit presented in Fig. 2.62 consists of elements with the following parameters: $V_{D,on} = V_{BE,on} = 0.7$ V, $V_{BE,\gamma} = 0.5$ V, $V_{CE,sat} = 0.2$ V, $\beta_F = 50$, $\beta_R = 0.1$.

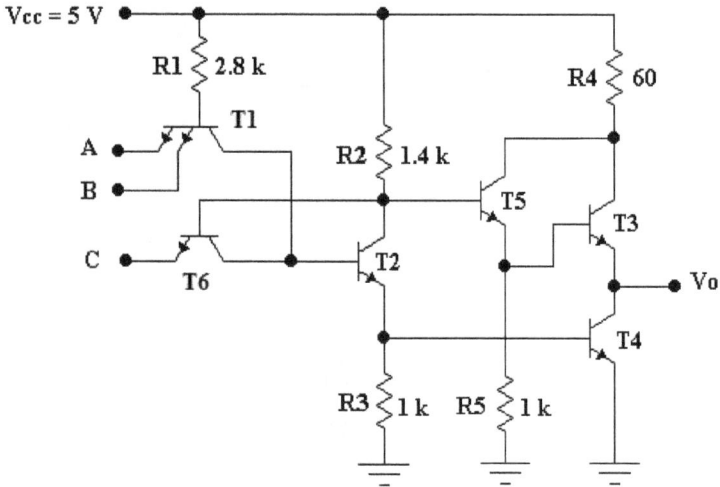

Fig. 2.62. Three-states TTL logic circuit.

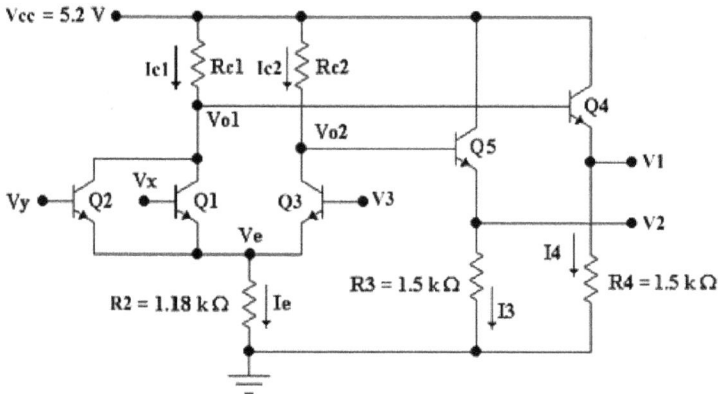

Fig. 2.63. ECL logic circuit.

a. Identify the logic function implemented by the presented circuit.

b. Calculate the maximum fan-out for the presented circuit.

2.8.10 The circuit presented in Fig. 2.63 consists of elements with the following parameters: $V_{D,on} = V_{BE,on} = 0.7$ V, $V_{BE,\gamma} = 0.5$ V, $V_{CE,sat} = 0.1$ V, $\beta_F = 50$, $\beta_R = 0.2$. Also, it is known that the voltage V3 = 4.15 V and the voltage of logic "1" is 4.5 V.

a. Identify the logic function implemented by the presented circuit.
b. Calculate the value of resistors Rc1 and Rc2.
c. Calculate the value of the logic "0" voltage.

Chapter 3

Logic Families Based on the Unipolar Devices

3.1. The Junction Field-Effect Transistor (JFET), a principle of operation

If we take a diode and attach to it the third electrode as shown in Fig. 3.1, we obtain a new controllable device with three electrodes, i.e. a transistor. However, it will be a specific transistor.

If we connect one electrode, source, to the ground and a second electrode, the drain, to the positive terminal of the external supply V_{DS}, a current begins to flow through the body of the device. Such a transistor is called a "normally open transistor". The current flows without a supply on the controlling electrode, gate. Evidently, two types of transistor may be built: the n-type transistor and the p-type transistor, according to the type of the conducting body. Let us consider the n-type transistor presented in Fig. 3.2.

Here, the electrical transport is carried out by charged carriers of one type, electrons, unlike in bipolar transistors. Therefore, this unipolar transistor has significantly reduced noises because the electrons do not have to cross a p-n junction while they are moving. The input resistance of this transistor is high, approximately 10^8 Ω, in contrast with the BJT transistor, which has a resistance of $\sim$50–200 Ω. As is known, a depletion region is situated between the n-type body and the p$^+$-type gate of this transistor, as in the usual diode. Due to the difference in impurity level between these two parts

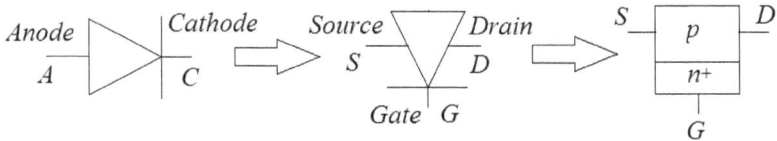

Fig. 3.1.   Three-electrode unipolar device.

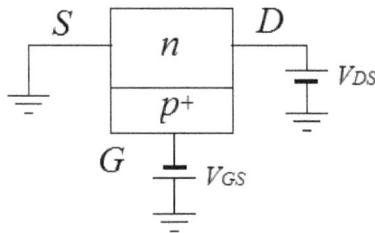

Fig. 3.2.   The normally open transistor.

of the transistor, the depletion region is basically in the n-type body. If we now begin to increase the negative supply on the gate, the width of the depletion region will increase and the current through the transistor between the drain and the source will decrease. The path for the current may be closed if enough high voltage is applied to the gate. So, in this case we can control the current flowing through the transistor using the gate's supply or using the electrical field applied to the depletion region of the transistor. Such a transistor is called the Junction Field-Effect Transistor (JFET). Let us consider its behavior quantitatively. Figure 3.3 illustrates the construction of a normally open JFET transistor. This is a non-planar construction, but it can be useful for analysis.

Transistor JFET, shown in Fig. 3.3, has a length of L and a width of W. The source of the device which is connected with the ground and the drain is supplied with the external supply V_{DS}. The body of the device, the n-type silicon, is located between two heavily doped regions of p^+-type silicon which play the role of the gate and which are supplied from the supply V_{GS}. The distance between the metallurgical borders of the p^+/n junctions is $2a$. The width of the depletion zone is designated as d. The current i_{DS} begins to flow from

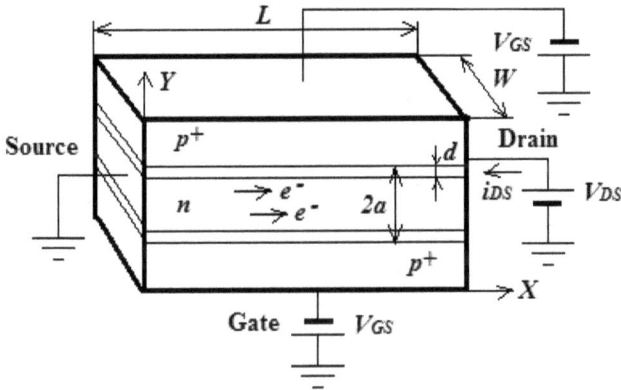

Fig. 3.3.　JFET transistor analysis.

the drain to the source when we connect the supply V_{DS} with the body of the transistor. The direction of movement of the electrons is shown in the picture. These electrons flow through the area equal to $A = W \times h$, where $h = 2(a - d)$ is the height of the conducting channel.

The height of the conducting channel, h, is a value which depends on the potential difference between the gate and the conducting channel. The gate is heavily doped and it represents an equipotential surface. The potential of the conducting channel is variable and depends on the voltage drop on the channel due to current flowing through the transistor. So, for a channel of n-type, we can apply the equation relating the width of the depletion region and the potential difference for the asymmetric p^+/n junction:

$$d(x) = \sqrt{\frac{2\varepsilon_0 \varepsilon_r (V_x - V_{GS})}{qN_D}} \tag{3.1}$$

where N_D is the impurity concentration in the body of the transistor and V_x is the voltage drop on the conductive channel over the distance x from the source. This voltage varies from V_{GS} at the source to $(V_{GS} - V_{DS})$ at the drain, as shown in Fig. 3.4. According to this equation, the width of the depletion region in both junctions, upper and lower, will be different, as shown in Fig. 3.4. The current density

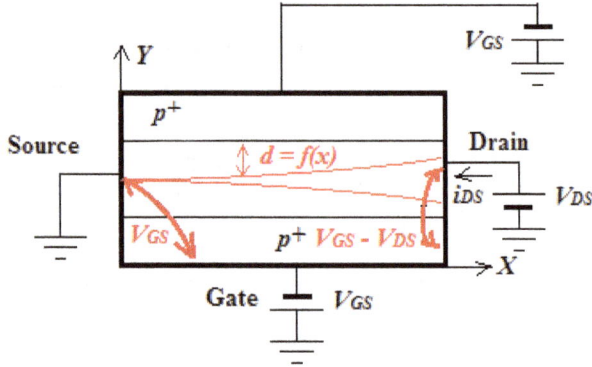

Fig. 3.4.   The distortion of the depletion region along the length of the transistor.

through the transistor is equal to the following:

$$j_{DS} = \sigma E(x) = -\sigma \frac{dV_x}{dx} = -qn\mu_e \frac{dV_x}{dx} \qquad (3.2)$$

where $\sigma = qn\mu_e$ is the conductivity of the conducting channel, μ_e is the mobility of the electrons, $E(x)$ is the electric field created by the supply V_{DS}, and n is the concentration of electrons.

A full current through the transistor can be calculated as follows:

$$i_{DS} = A \cdot j_{DS} = 2(a - d)W \cdot j_{DS}. \qquad (3.3)$$

If we take into account that $n = N_D$ and substitute values d and j_{DS} into Eq. (3.3), we will obtain the following equation:

$$i_{DS} = -2WqN_D\mu_e \left\{ a - \left[\frac{2\varepsilon_0\varepsilon_r}{qN_D}(V_x - V_{GS}) \right]^{0.5} \right\} \frac{dV_x}{dx}. \qquad (3.4)$$

We can obtain the solution for this equation after the integration of the following equations:

$$\int_0^L i_{DS}dx = -2WqN_D\mu_e \int_0^{V_{DS}} \left\{ a - \left[\frac{2\varepsilon_0\varepsilon_r}{qN_D}(V_x - V_{GS}) \right]^{0.5} \right\} dV. \qquad (3.5)$$

$$i_{DS} = -\frac{2WqaN_D\mu_e}{L}$$

$$\times \left\{ V_{DS} - \frac{2}{3}\left(\frac{2\varepsilon_0\varepsilon_r}{qN_Da^2}\right)^{0.5}[(V_{DS} - V_{GS})^{1.5} - (-V_{GS})^{1.5}]\right\}.$$

$$(3.6)$$

In this equation, the coefficient placed before the part of the equation enclosed within braces is a constant value which has a dimension of conductivity:

$$G_0 \equiv \frac{2WqaN_D\mu_e}{L}.$$

$$(3.7)$$

It is constant and defined by the type of material (μ_e), the technological parameters (a, N_D) of the material, and the geometric dimensions (L, W) of the material.

If we return to Eq. (3.1), we can calculate the maximum value of the voltage which disables the current flow through the transistor. Assuming that $d_{max}(x) = a$, and squaring both sides of Eq. (3.1), we obtain the following equation:

$$a^2 = \frac{2\varepsilon_0\varepsilon_r}{qN_D}(V_x - V_{GS}).$$

$$(3.8)$$

Now, we can define the voltage which stops the current from passing through the transistor, or the **pinch-off voltage**, as follows:

$$V_p \equiv V_{GS} - V_x = -\frac{qN_Da^2}{2\varepsilon_0\varepsilon_r}.$$

$$(3.9)$$

After that, Eq. (3.6) takes the following form:

$$i_{DS} = -G_0\left\{V_{DS} - \frac{2}{3\sqrt{V_p}}[(V_{DS} - V_{GS})^{1.5} - (-V_{GS})^{1.5}]\right\}.$$

$$(3.10)$$

An analysis of this equation shows that if the voltage V_{DS} supplied to the body of the transistor is low, the equation reduces to the following:

$$i_{DS} = -G_0V_{DS}.$$

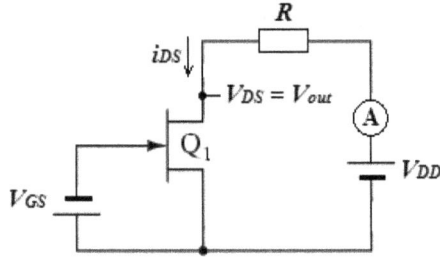

Fig. 3.5.   The simplest electrical scheme connecting the JFET transistor of n-type.

i.e. the current is directly proportional to the applied voltage. The transistor works in the Ohmic mode and represents a variable resistor. Let us consider the simplest electrical connecting scheme for the JFET transistor, see Fig. 3.5.

Assume that we provide measurements of the I–V characteristics of the JFET transistor presented in Fig. 3.5. Without the input voltage, voltage V_{GS}, the transistor conducts the full current i_{DS} and begins to work as a variable resistor with low V_{DS}. With the growth of V_{DS}, the I–V characteristic changes its shape due to changes in the potential difference between the conducting channel and the gate electrode along the length of the transistor. When the V_{DS} reaches the value $(V_{GS} - V_p)$, the transistor switches to the saturation mode and the current i_{DS} does not increase further. The substitution of the equation $V_{DS} = V_{GS} - V_p$ into Eq. (3.10) shows that the current i_{DS} is not dependent on V_{DS} in the saturation mode:

$$i_{DS(sat)} = -G_0 \left[V_{GS} - \frac{2(-V_{GS})^{1.5}}{3 \cdot \sqrt{V_p}} - \frac{V_p}{3} \right]. \qquad (3.11)$$

The increasing of the input voltage, V_{GS}, enables us to build several other I–V characteristics so that together they complete the family of I–V characteristics for the JFET transistor. This family of I–V characteristics is shown in Fig. 3.6.

It should be noted that the input voltage values shown in the figure are taken at random and are just examples. However, the regions corresponding to the modes of work — the Ohmic mode and the saturation mode — are shown here. The relation between

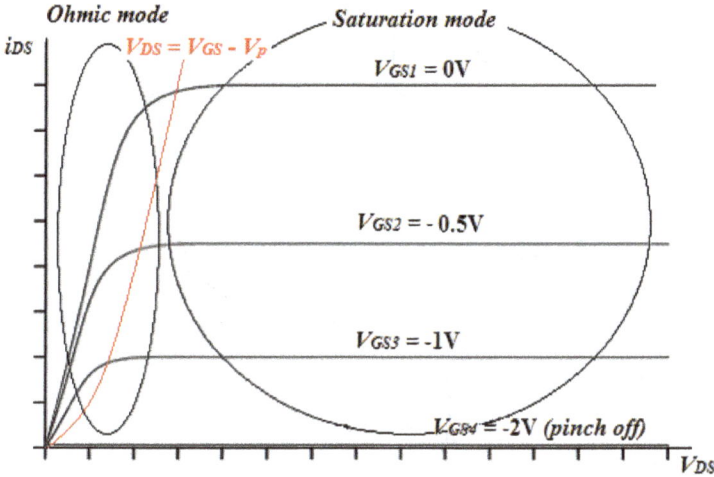

Fig. 3.6. The I–V characteristics family for the JFET transistor.

different drain currents in the saturation mode is parabolic:

$$i_{DS(sat)} = I_{DSS} \left(1 - \frac{V_{GS}}{V_p} \right)^2 \tag{3.12}$$

where I_{DSS} is the maximum drain current measured on condition that $V_{GS} = 0$.

The usual JFET transistors are not widely used in digital circuits, but they serve as a basis for digital circuits employing MESFET transistors.

3.2. The MOSFET (IGFET) transistor

3.2.1. *The enhancement-type NMOS transistor, a principle of operation*

Let us consider a construction built using planar technology, i.e. grown only on one side of the silicon wafer, using the series of consecutive operations which was presented in Fig. 3.7. As shown in the picture, the metal-oxide-semiconductor field effect transistor (MOSFET) or the insulated-gate field effect transistor (IGFET) consists of a thin layer system grown on the surface of the silicon

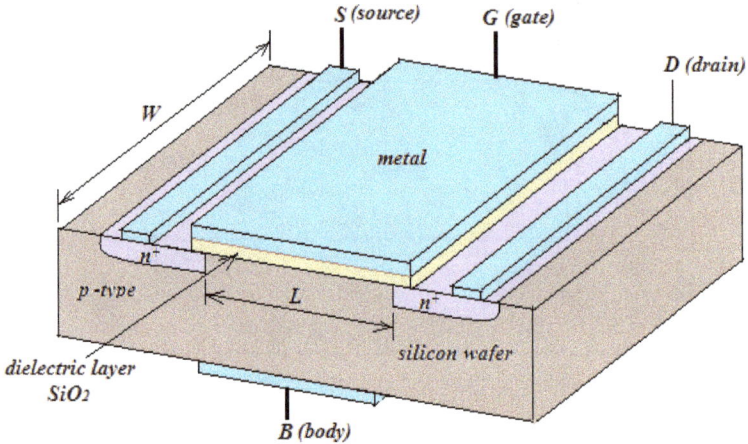

Fig. 3.7. External view of a MOSFET transistor of n-type (NMOS).

wafer of p-type. The thickness of the wafer usually equals to 200–300 μm. The depth of the heavily doped regions used for the connection of the source and the drain does not exceed 0.5 μm. These heavily doped regions, usually called wells or pockets, are built using diffusion or ion implantation technologies.

The length, L, and the width, W, of the transistor are defined by the technological processes using the transistor for growth and may vary from several tens of nanometers up to several microns. All the electrodes in the system may be built from specific metals or alloys. Sometimes, a very-conductive poly-crystalline silicon is applied to achieve this goal. The thickness of the metal electrodes and connection lines is approximately 200 nm; however, it is defined by the density of the current conducted through these lines. A dielectric layer isolating the gate electrode from the body of the transistor is usually made from silicon dioxide, or SiO_2, a glasslike substance grown on silicon using various oxidation technologies. The thickness of the dielectric layer is equal to 20–100 nm. Actually, the process of transistor manufacturing includes a whole range of scientific and technological activities which are not addressed in this course.

Usually, the source and body of the transistor are connected to the ground, the drain is supplied with the external supply V_{DD}, and

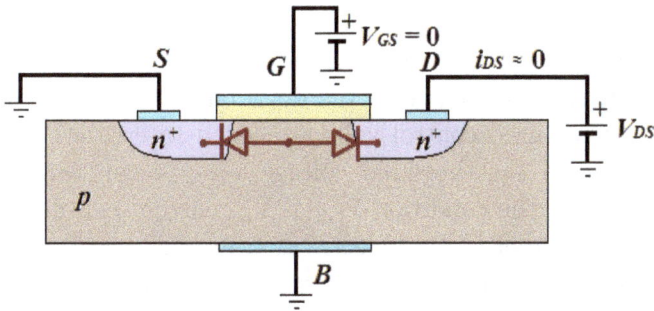

Fig. 3.8. The state of the transistor without the control voltage.

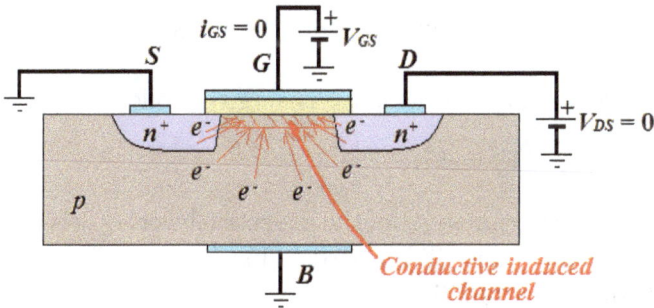

Fig. 3.9. The formation of the inductive induced channel in the NMOS transistor.

the control electrode, the gate, is connected to the control supply V_{GS}. When a control voltage $V_{GS} = 0$, regardless of the voltage V_{DS}, current cannot flow between the drain and the source as both the source-body and body-drain junctions represent two oppositely connected diodes, as shown in Fig. 3.8. So, the maximum possible current i_{DS} is the leakage current of a diode which does not exceed $\sim 10^{-12}$ A. Because of that, this transistor is called a normally closed transistor.

Assume that the power supply $V_{DS} = 0$ and V_{GS} is positive with respect to the source. This positive potential then begins to attract electrons from heavily doped regions and from the body when electrons are minority carriers as shown in Fig. 3.9.

This process continues with the increase of voltage V_{GS} until the conductive inverse layer, called an induced layer, is created with an electron concentration equal to the body basic concentration p. The value of the required control voltage needed to create this induced channel is called the threshold voltage V_T. We can say that after compliance with the condition $V_{GS} > V_T$, current can flow through the induced channel and all electrical transport between the drain and the source is carried out by one type of charged carrier, electrons in our case. Therefore, we are talking about a **unipolar** transistor. It should be emphasized that in the static state, the control current $i_{GS} = 0$ always since the dielectric layer does not allow any current. Therefore, all control of the current flow between the drain and the source is carried out using the potential V_{GS} with respect to the source.

The value V_T may be calculated using the following equation:

$$V_T = V_{MS} - \frac{Q_{ox}}{C_{ox}} + \frac{2\emptyset_F}{q} - \frac{Q_b}{C_{ox}} \tag{3.13}$$

where C_{ox} is the permittivity of the dielectric layer, Q_{ox} is the specific charge on the dielectric-silicon border, V_{MS} is the difference between the work functions of silicon and the electrode material, Q_b is the charge associated with the negative ions remaining in the depleted zone of the silicon after leaving its mobile holes, and $\emptyset_F = E_{Fi} - E_F$ is the difference between the Fermi levels of the intrinsic and doped silicon.

Figure 3.10 illustrates the state of the transistor with $V_{GS} > V_T$ and V_{DS} low.

In this state, when the voltage V_{DS} is low, the current i_{DS} will be proportional to the voltage V_{DS}. This mode of work is called the triode or Ohmic mode. One can say that the transistor behaves as a linear resistor in this mode. When the voltage V_{DS} continues to rise, a distortion effect similar to that in the JFET transistor will occur. Due to the different potential differences between the gate and various points in the induced channel, as shown in Fig. 3.11, the shape of the induced layer will change and we obtain the so-called pinch-off effect. When reaching $V_{GS} - V_{DS} = V_T$, the transistor goes into the saturation mode.

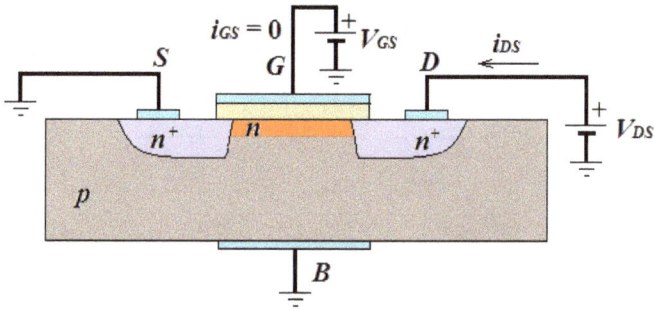

Fig. 3.10. The NMOS transistor with an induced layer.

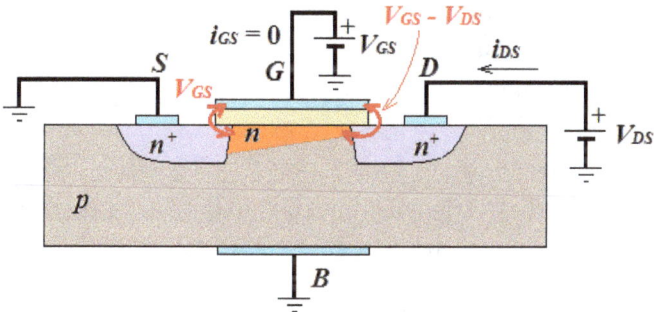

Fig. 3.11. The distortion effect of the induced channel shape under $V_{DS} > V_{GS} - V_T$.

As shown in Fig. 3.11, this distortion of the conducting layer's shape is caused by a difference between the voltages at both ends of the gate: V_{GS} at the source's end and $(V_{GS} - V_{DS})$ at the drain's end. For a quantitative description of the transistor's behavior, let us look at Fig. 3.12.

We will consider a small part of the conductive channel with length dx which is at a distance x from the source. Due to voltage drop along the channel, the voltage between the gate and point x of the channel will be $(V_{GS} - V_x)$. If we take into account that a channel only begins conducting current on condition that $V_{GS} > V_T$, we can calculate an elementary charge of electrons dq moving from the source to the drain:

$$dq = -C_{ox} W dx (V_{GS} - V_x - V_T). \qquad (3.14)$$

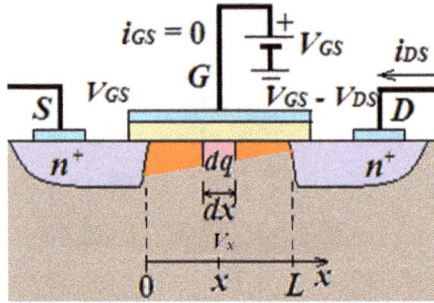

Fig. 3.12.   The derivation of the NMOS behavior equation.

Here, dq is the charge accumulated in the virtual plane capacitor built with the gate and channel. Accordingly, the C_{ox} is the specific capacity of this capacitor:

$$C_{ox} = \frac{\varepsilon_{ox}}{d_{ox}} \qquad (3.15)$$

where ε_{ox} is the dielectric constant of the applied dielectric material (in our case it is silicon dioxide, SiO_2) and d_{ox} is its thickness.

Our elementary charge moves under the influence of an electric field produced by the applied voltage. The speed of the charge is proportional to the electric field:

$$\frac{dx}{dt} = -\mu_e E(x) = \mu_e \frac{dV_x}{dx}. \qquad (3.16)$$

The current will be equal to the product of the elementary charge and its moving speed:

$$i = dq \times \frac{dx}{dt} = -\mu_e C_{ox} W (V_{GS} - V_x - V_T) \frac{dV_x}{dx}. \qquad (3.17)$$

If we divide the variables and integrate this equation over the length of the transistor, we obtain the current i_{DS}:

$$i_{DS} = \int_0^L i\, dx = -\int_0^{V_{DS}} \mu_e C_{ox} W (V_{GS} - V_x - V_T) dV_x. \qquad (3.18)$$

$$i_{DS} = \mu_e C_{ox} \left(\frac{W}{L}\right) \left[(V_{GS} - V_T) V_{DS} - \frac{1}{2} V_{DS}^2 \right]. \qquad (3.19)$$

Equation (3.19) describes the behavior of the NMOS transistor in the triode mode of the work. To describe the behavior of the transistor in the saturation mode, we need to substitute the saturation condition $V_{DS} = V_{GS} - V_T$ into Eq. (3.19):

$$i_{DS} = \frac{1}{2} \mu_e C_{ox} \left(\frac{W}{L} \right) (V_{GS} - V_T)^2 \qquad (3.20)$$

where the current i_{DS} does not depend on the voltage V_{DS} and represents the parabolic function of the gate voltage V_{GS}. Here, the product $K'_n = \mu_n C_{ox}$ is called the process transconductance parameter and the product $K_n = K'_n (W/L)_n$ is called the transconductance of the NMOS transistor.

Let us consider the simplest electrical connecting scheme for the NMOS transistor, see Fig. 3.13. Let us assume that we provide measurements of the I–V characteristics of the NMOS transistor which is presented in Fig. 3.13. Without the input voltage, voltage V_{GS}, the transistor will not conduct the current i_{DS} and will continue to be closed up. The inverse conductive layer will not be created until the input voltage rises up to the value of the voltage threshold, $V_{GS} = V_T$.

When the input voltage is more than the V_T value, the collector current i_{DS} begins to flow through the transistor and for each value of $V_{GS} > V_T$ we can measure this current. At low V_{DS}, the current changes in a linear manner that is described by Eq. (3.19) if we

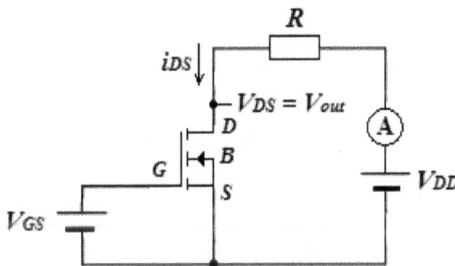

Fig. 3.13.   Simplest electrical scheme of connecting the NMOS transistor.

assume that $V_{DS} \approx 0$:

$$i_{DS} = K_n(V_{GS} - V_T)V_{DS}. \qquad (3.21)$$

On the condition that the transistor works as a variable resistor, we designate this working mode as the triode mode. In addition, we can calculate the approximate value of our non-linear resistor using Eq. (3.21):

$$R_{DS} = \frac{V_{DS}}{i_{DS}} = \frac{1}{K_n(V_{GS} - V_T)}. \qquad (3.22)$$

Evidently, the minimum value of the resistance designated as R_{on} will be reached when $V_{GS} = V_{DD}$:

$$R_{on} = \frac{1}{K_n(V_{DD} - V_T)}. \qquad (3.23)$$

With the growth of V_{DS}, the I–V characteristic changes its shape due to changes in the potential difference between the conducting channel and the gate electrode along the length of the transistor. When the V_{DS} reaches the value $(V_{GS} - V_T)$, the transistor switches to the saturation mode and the current i_{DS} does not increase further. This behavior is described by Eq. (3.20).

Increasing the input voltage, V_{GS}, enables us to build several other I–V characteristics so that together they complete the family of I–V characteristics for the NMOS transistor. This family of I–V characteristics is shown in Fig. 3.14. As shown in this figure, there are three different modes of operation for the NMOS transistor defined by the following relations:

1. Cut off mode — $V_{GS} \leq V_T$.
2. Triode mode — $V_{DS} < V_{GS} - V_T$.
3. Saturation mode — $V_{DS} \geq V_{GS} - V_T$.

3.2.2. *The depletion-type NMOS transistor, a principle of operation*

Let us consider a transistor with a built-in conductive layer of n-type, as shown in Fig. 3.15. Such a conductive layer may be easily grown

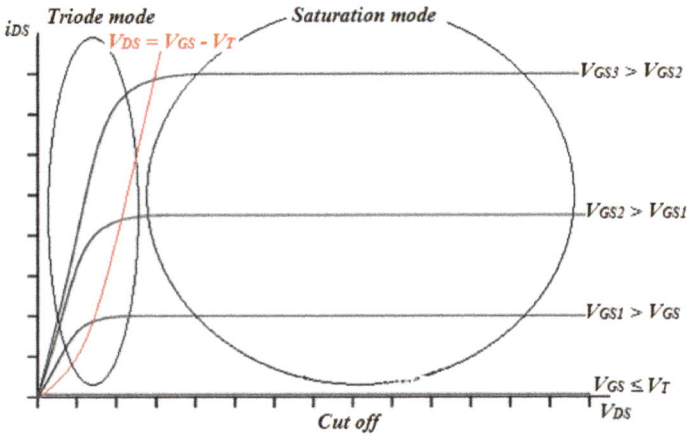

Fig. 3.14. The I–V characteristics family for the NMOS transistor.

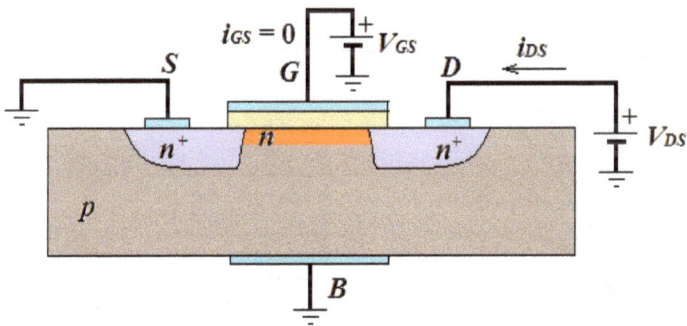

Fig. 3.15. The depletion-mode transistor.

during the transistor preparation process. The density of donors in this case should be equal to the density of acceptors in the wafer.

The transistor which is shown in Fig. 3.15 is called a depletion-mode transistor or a "normally open transistor." Such a transistor can conduct a current without any voltage on the gate. Moreover, to close the transistor, we need to apply a negative voltage equal to the V_T to the gate.

The input characteristics of the depletion-mode transistor are presented in Fig. 3.16.

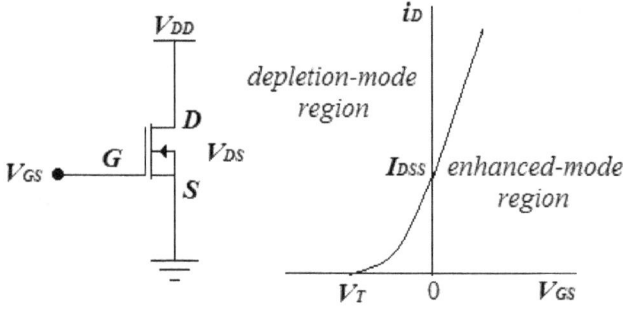

Fig. 3.16.   The input characteristics of the depletion NMOS transistor.

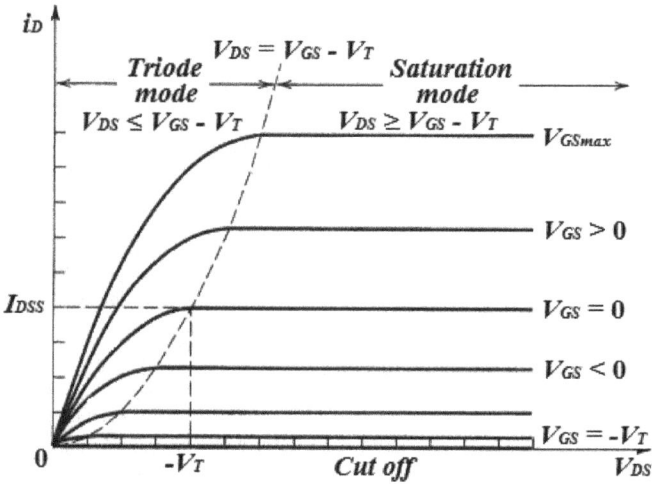

Fig. 3.17.   Output characteristics of the depletion-mode transistor.

As shown, this characteristic represents a diode characteristic shifted at the negative input voltages. The depletion-mode transistor can work at two regions of the input voltage and the current through the transistor flows without the voltage applied to the gate. The current i_D flowing through the transistor on condition that $V_{in} = 0$ is called I_{DSS} — a basic saturation current. Figure 3.17 represents a family of output characteristics for the depletion-mode NMOS transistor. This transistor can be applied in many ways and can be flexibly used in different devices. For example, in linear amplifiers,

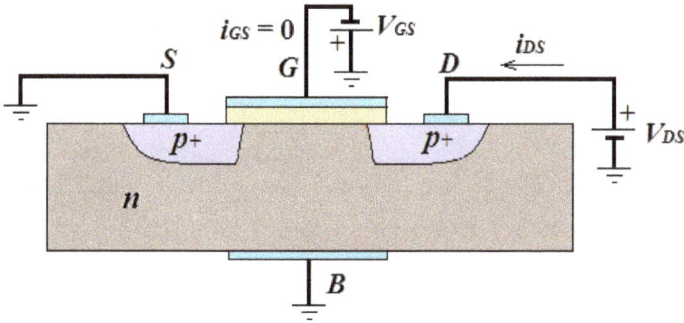

Fig. 3.18.   PMOS transistor construction.

this transistor does not require additional bias on the gate to make it applicable without the distortion of the input signal.

3.2.3.   *The PMOS transistor, a principle of operation*

Let us consider a field effect transistor grown on an n-type silicon wafer and built to resemble the NMOS transistor as shown in Fig. 3.18.

The main difference between this transistor and the NMOS enhanced mode transistor is the substrate silicon of n-type and the two heavily doped $p+$ regions designated as the source and the drain. Now, if we apply a negative voltage to the gate electrode, an inverse layer of p-type conductivity will be built on the border between the gate's insulator and the substrate. This layer represents the conductive p-type channel connecting the source and drain regions and enabling a current i_D to pass through the transistor. Such types of transistors are called PMOS transistors due to the type of the inverse layer. As the mobile charged carriers in this case are holes, the switching rate of these transistors is less than that of NMOS transistors. PMOS transistors may be of two types, enhanced mode or depletion mode. Usually, they are of enhanced mode.

Figure 3.19 shows a designation of a PMOS transistor of enhanced mode and the simplest electrical switching scheme using this transistor.

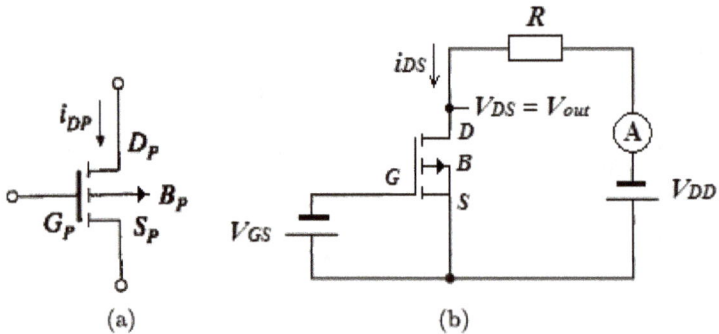

Fig. 3.19. (a) PMOS transistor and (b) the simplest electrical scheme using a PMOS transistor.

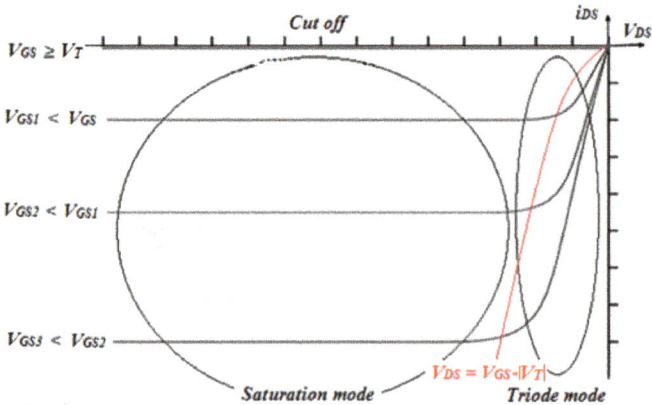

Fig. 3.20. Current-voltage characteristic of the PMOS transistor.

As shown, the designation of the PMOS transistor has an arrow directed from the gate of the transistor to the body. A PMOS transistor requires a negative voltage on the gate to switch actions. This voltage should be more negative than the threshold voltage V_t. Accordingly, all control voltages and currents in the transistor will be negative. An I–V characteristic of the PMOS transistor is shown in Fig. 3.20 for various voltages on the gate, V_{GS}.

The PMOS transistor is similar to the NMOS transistor in that we can identify three regions of transistor operation. As shown in

Fig. 3.20, there are three different modes of operation for the NMOS transistor defined by the following relations:

1. Cut off mode: $V_{GS} \geq V_T$; $i_{DS} = 0$.
2. Triode mode: $0 > V_{DS} \geq V_{GS} - V_T$; $i_{DS} = -\mu_e C_{ox}(\frac{W}{L})[(V_{GS} - V_T)V_{DS} - \frac{1}{2}V_{DS}^2]$.
3. Saturation mode: $V_{DS} \leq V_{GS} - V_T < 0$; $i_{DS} = -\frac{1}{2}\mu_e C_{ox}(\frac{W}{L})(V_{GS} - V_T)^2$.

The PMOS transistors together with the NMOS transistors may be designated in reference sources or in electrical schemes in various ways. They may be designated as four-terminal devices consisting of source (S), gate (G), drain (D), and body (B) electrodes or as three-terminal devices, without the body electrode. In this case it means that the body electrode is connected to the source by the manufacturer. Figure 3.21 represents four types of designation for various types of MOS transistors: PMOS and NMOS, four-terminals and three-terminals, enhanced and depletion modes.

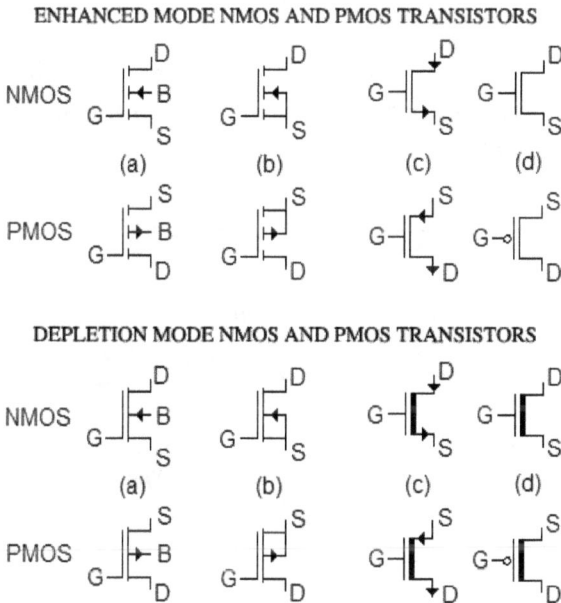

Fig. 3.21.   The designations of various MOS transistors in the schemes.

3.3.   N-type metal-oxide-semiconductor (NMOS) logic

3.3.1.   *The NMOS inverter with resistive load*

Let us consider the electrical circuit presented in Fig. 3.22. This circuit consists of two electronic devices: an NMOS transistor and a constant resistor R_D.

Usually, this circuit is a little part of the large net of electronic devices. As each logic circuit works in the environment of the logic circuits related to the same logic family, the load for the circuit presented in Fig. 3.22 is the same circuit or a number of the same logic circuits connected in parallel. Therefore, such a circuit is loaded with the same circuit, which has an NMOS transistor in the input. A chain of NMOS inverters is shown in Fig. 3.23.

The input resistance of this circuit is very large, it is equal to $\sim 10^{12}\Omega$ and it represents the input capacity of the NMOS transistor. So, we can present the load of the circuit by an equivalent capacitor C_L, as shown in Fig. 3.23. It follows that the input current will be zero at steady state. Therefore, an input current and an output current will only be present in this circuit during the transition process from logic "0" to logic "1" or vice versa. In the other cases where an output current exists, the output end is connected to a resistor or lamp or something else. They are unusual.

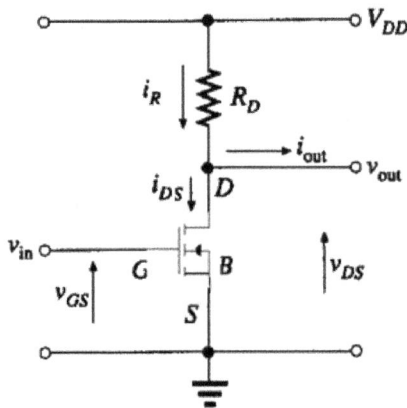

Fig. 3.22.   A simple inverter based on the NMOS transistor.

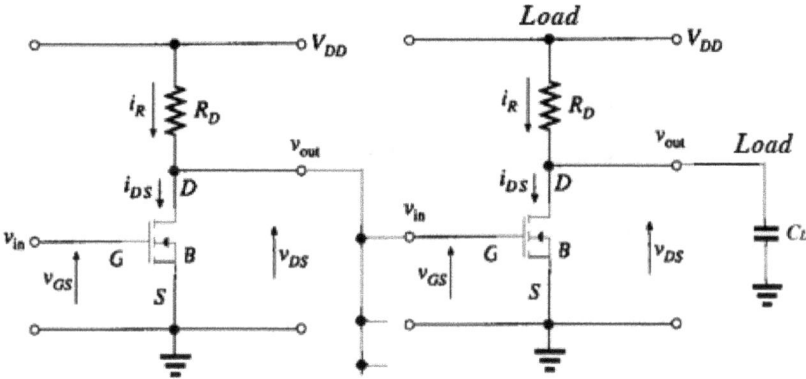

Fig. 3.23.   A loaded NMOS inverter.

To analyze the circuit presented in Fig. 3.22 we should return to the usual method described in the second chapter. The main function of this circuit may be defined by the state of the output electrode for different values of the input voltage. When the input voltage is zero, the transistor will be in the cutoff state. So, due to the capacitive load, one can say that the current passing through the transistor will be zero also. In this case, the output voltage will be equal to the supply voltage $V_{out} = V_{DD}$, which represents the logic "1".

If the input voltage is equal to V_{in} = "1" = V_{DD}, the NMOS transistor will be in the state enabling the transfer of the maximum current. To estimate this current and analyze the circuit behavior, let us consider the I–V characteristics of our circuit which is presented in Fig. 3.24.

This figure represents a geometric assessment of the behavior of the above circuit. The circuit consists of two elements: an NMOS transistor and a resistor. The transistor's behavior is described by its I–V characteristic family. The current flowing through the resistor is represented by the blue line AD and described by the following equation:

$$V_{DD} = i_R R_D + V_{DS} = i_R R_D + V_{out}. \qquad (3.24)$$

As both currents should be equal, $i_R = i_{DS}$, the line segment AC represents the locus of points of exact solutions for Eq. (3.24)

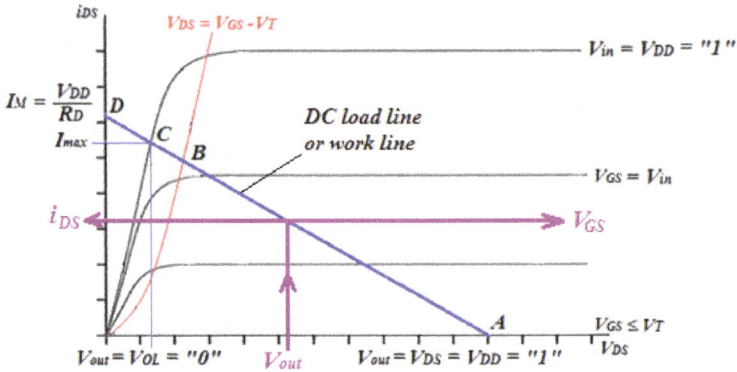

Fig. 3.24.   The I–V characteristic of the NMOS inverter.

together with the transistor's equations. This segment AC is called the work line of the circuit. Point A of this segment represents the cutoff behavior of the transistor when the current $i_{DS} = 0$ and when the output voltage is in the state of logic "1", $V_{out} = V_{DD}$. Point D represents the maximum possible current though the resistor assuming that the on-resistance of the transistor is zero, $I_M = V_{DD}/R_D$. Point C represents the real maximum current through the resistor and transistor, I_{max}, on condition that $V_{in} = V_{DD} = $ "1" is the maximum input voltage. The output voltage at this point, C, will be the minimum possible output voltage $V_{out} = V_{OL}$ in this circuit. This voltage may be designated as the logic "0", $V_{OL} = $ "0". We saw that at the minimum input voltage the circuit is set to logic "1", and at the maximum input voltage, the output is a logic "0", so our circuit is an inverter. It implements the logic function NOT and during the operation is switched from point A to point C, and vice versa.

Point B represents a transition point between the triode and saturation regions of operation of the transistor, and is described with the equation $V_{DS} = V_{GS} - V_t$. As the location of point C is left from point B, a current flowing through the transistor at point C is described by Eq. (3.19), which is related to the triode region. Using this equation, one can calculate the lower output voltage $V_{out} = V_{OL} = $ "0". To achieve this goal we will combine Eqs. (3.19) and (3.24):

$$V_{out} = V_{DD} - R_D i_{DS} = V_{DD} - \frac{K_n}{2} R_D [2(V_{GS} - V_T)V_{DS} - V_{DS}^2].$$
(3.25)

At point C, the input voltage is equal to logic "1", therefore

$$V_{OL} = V_{DD} - \frac{K_n}{2} R_D [2(V_{DD} - V_T)V_{OL} - V_{OL}^2].$$
(3.26)

If we neglect the second order of smallness, the approximate solution of this equation will be presented as follows:

$$V_{OL} = \frac{V_{DD}}{1 + K_n R_D (V_{DD} - V_T)}.$$
(3.27)

Now, when the extreme operating points are defined on the work line using the I–V characteristics of the inverter (see Fig. 3.24), we can build a transfer function describing the transition behavior of the NMOS inverter. To achieve this goal, we will divide the transition performance during the switch from logic "1" to logic "0" into three steps defined by the transistor behavior mode:

1. Cut off mode, $V_{in} = V_{GS} \leq V_T$, $i_{DS} = 0$, $V_{out} = $ "1", the O-A region.
2. Saturation mode, $V_{in} \leq V_{DS} + V_T = V_{out} + V_T$, $i_{DS} = \frac{1}{2}\mu_e C_{ox}$ $(\frac{W}{L})(V_{GS} - V_T)^2$, the A-B region.
3. Triode mode, $V_{in} > V_{DS} + V_T = V_{out} + V_T$, $i_{DS} = \mu_e C_{ox}(\frac{W}{L})[(V_{GS} - V_T)V_{DS} - \frac{1}{2}V_{DS}^2]$, the B-C region.

Now, we can build the transfer function. In the first region, O-A, the function is constant and equal to $V_{DD} = V_{OH} = $ "1".

In the second region, A-B, we can obtain the function behavior from a combination of Eqs. (3.20) and (3.24).

$$V_{out} = V_{DD} - R_D i_{DS} = V_{DD} - \frac{K_n}{2} R_D (V_{GS} - V_T)^2$$

$$= V_{DD} - \frac{K_n R_D}{2}(V_{in} - V_T)^2.$$
(3.28)

As shown, this function represents a branch of an inverted parabola with vertex coordinates $V_{in} = V_T$ and $V_{out} = V_{DD}$, (V_T, V_{DD}). This parabola finishes its propagation at point B. Here, the transistor changes its mode of operation from the saturation mode to the triode mode. At point B, $V_{outB} = V_{DS} = V_{GS} - V_T = V_{inB} - V_T$.

We obtain the equation describing the behavior of the transfer function in the third region from Eq. (3.25):

$$V_{out} = V_{DD} - R_D i_{DS}$$

$$= V_{DD} - \frac{K_n}{2} R_D [2(V_{GS} - V_T)V_{DS} - V_{DS}^2]$$

$$= V_{DD} - \frac{K_n R_D}{2} [2(V_{in} - V_T)V_{out} - V_{out}^2]. \qquad (3.29)$$

This equation can be solved for the input voltage, V_{in}:

$$V_{in} = \frac{V_{DD}}{V_{out}K_n R_d} - \frac{V_{out}(V_T K_n R_D - 1)}{V_{out}K_n R_D} + \frac{V_{out}}{2}. \qquad (3.30)$$

Where the first term of the sum is a hyperbolic function, the second term is a constant and the third term is the proportional summand. So, the transfer function is hyperbolic in the region B-C. Now, we can connect the points B and C with a hyperbolic curve. Evidently, our curve is drawn approximately. Figure 3.25 represents the transfer function which we built.

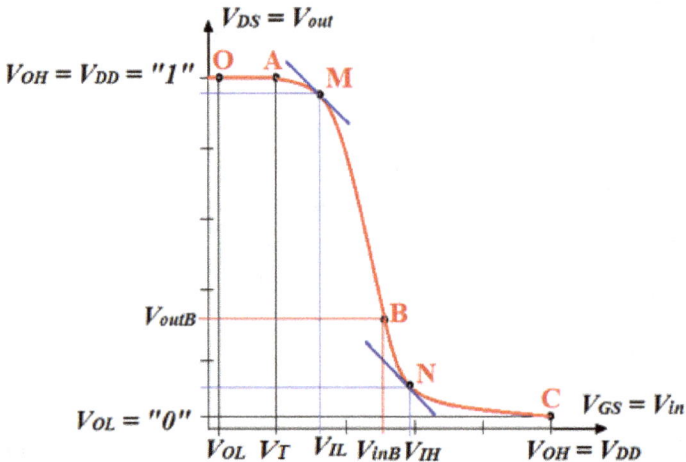

Fig. 3.25.   The transfer function of the NMOS inverter with the resistive load.

This curve reflects the behavior of the NMOS inverter when switching from one logic state to another.

To characterize the stability of this inverter in the presence of noise or input voltage fluctuations, we should estimate the value of the noise margins. In other words, we should calculate the coordinates of the points M and N when the derivative of the transfer function is equal to -1. For each point, our calculations will be different, as the point M belongs to the saturation mode of the transistor and the point N belongs to the triode mode of operation.

We should calculate the coordinates of the point M in the following way. Firstly, we will take Eq. (3.28) and differentiate it:

$$V_{out} = V_{DD} - \frac{K_n R_D}{2}(V_{in} - V_T)^2$$

$$\frac{dV_{out}}{dV_{in}} = -K_n R_D(V_{in} - V_T) = -1. \qquad (3.31)$$

From this equation we will obtain the input voltage at point M:

$$V_{IL} = V_{inM} = V_T + \frac{1}{K_n R_D}. \qquad (3.32)$$

If we take the value of the input voltage at point M and substitute in the initial Eq. (3.28), we obtain the value of V_{outM}:

$$V_{outM} = V_{DD} - \frac{1}{2K_n R_D}. \qquad (3.33)$$

To calculate the coordinates of the point N and the input voltage V_{IH}, we need to repeat the derivation for Eq. (3.30). After some simple transformations, we obtain the required values:

$$V_{IH} = V_{inN} = V_T - \frac{1}{K_n R_D} + \sqrt{\frac{2V_{DD}}{3K_n R_D}}. \qquad (3.34)$$

$$V_{outN} = \sqrt{\frac{2V_{DD}}{3K_n R_D}}. \qquad (3.35)$$

Now we can find the noise margins for the NMOS inverter using Eqs. (1.4):

$$NM_L = V_{IL} - V_{OL} \quad and \quad NM_H = V_{OH} - V_{IH}.$$

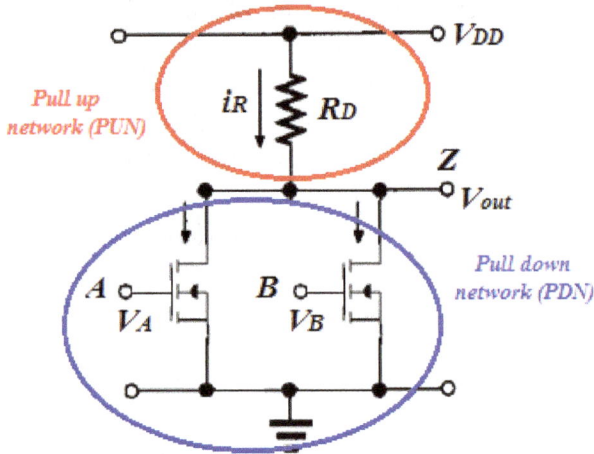

Fig. 3.26. A logic gate NOR based on the NMOS transistors.

If we designate the derivation of the transfer function as a magnification coefficient

$$A_V = \frac{dV_{out}}{dV_{in}} \tag{3.36}$$

we can say that the inverter is stable when $A_V < |1|$ and is not determined from the logic point of view in other cases.

The NMOS inverter considered above implements the logical function "NOT". As is known, to complete Boolean algebra we also need to realize the logical functions "OR" and "AND". Figure 3.26 represents a possible logical gate built using two NMOS transistors connected in parallel. If we compare this circuit with the circuit from Fig. 3.22, we can see that each circuit consists of two different parts, as shown in Fig. 3.26: the upper part called a pull up network (PUN) and the lower part called a pull down network (PDN). The PUN consists of all the elements of the circuit which exist between two lines: the supply line connected with the supply V_{DD} and the output line. The PDN consists of all the elements of the circuit which exist between two lines: the output line and the ground line. Both parts may contain various electron elements. Our NMOS transistors are in the PDN part, therefore they will be conductive when the input voltages V_A and V_B are high, or equal to V_{DD}. The output point

Z will be connected with the ground if transistors are in the triode mode of operation.

To analyze the circuit presented in Fig. 3.26, we need to build a truth table. The inputs A and B should take all possible logic states and the output voltage should be evaluated in all cases. Evidently, in our construction, the high voltage in one or both of our transistors reduces the output voltage. Only a low input signal from both input electrodes A and B will bring the output voltage to the high state, $V_{out} = $ "1". So, the truth table will look as follows:

N	A	B	V_0
1	0	0	1
2	0	1	0
3	1	0	0
4	1	1	0

As seen, this table corresponds with the logic function NOR. Another type of transistor connection gives us the function NAND: only if both inputs are high will the output voltage be low. This gate is shown in Fig. 3.27.

The truth table for the gate presented in Fig. 3.27 will look as follows:

N	A	B	V_0
1	0	0	1
2	0	1	1
3	1	0	1
4	1	1	0

The disadvantages of NMOS logic gates with resistive loads include:

- A highly asymmetric transfer function that leads to asymmetric noise margins;

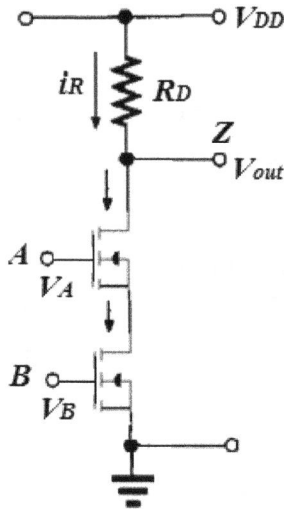

Fig. 3.27.   A logic gate NAND based on the NMOS transistors.

- A relatively broad gain region $(V_{IH} - V_{IL})$ when a logical gate is out of logic;
- The enormous size of resistors in comparison with transistors, making it difficult to seal the electronic circuitry;
- High dissipation power in the static logic states, as the current flows through the circuit in the low logic state.

3.3.2.   *The NMOS inverter with NMOS load of the enhancement type*

Let us consider an enhanced mode NMOS transistor with a shortly connected gate and drain, see Fig. 3.28. This construction is called a diode-connected transistor.

Due to the connection, $V_{GS} = V_{DS}$ always. Evidently, such a transistor will close when $V_{GS} = V_{DS} \leq V_T$. A current through this transistor will begin at $V_{DS} > V_T$ only. As $V_{DS} > V_{GS} - V_T$ always, the diode-connected enhanced mode NMOS transistor will have two states only: cutoff at $V_{GS} = V_{DS} \leq V_T$ and saturation at $V_{GS} = V_{DS} > V_T$. Figure 3.29 represents an I–V characteristic of the diode-connected transistor.

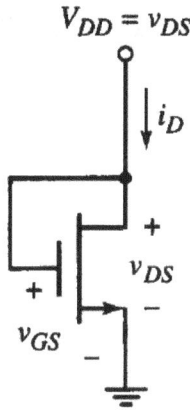

Fig. 3.28. A diode-connected enhanced mode NMOS transistor.

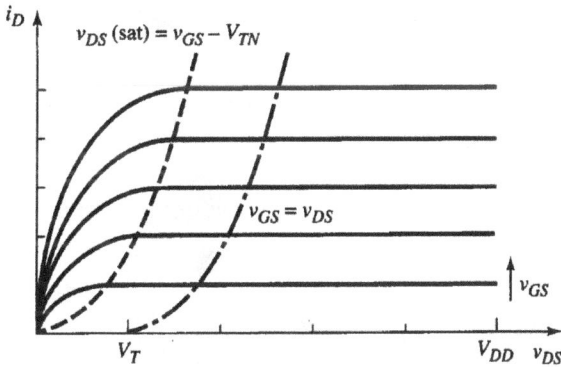

Fig. 3.29. The I–V characteristic of the diode-connected NMOS transistor.

A current flowing through the transistor, as may be seen in Fig. 3.29, represents a parabolic behavior

$$i_{DS} = \frac{1}{2}\mu_e C_{ox}\left(\frac{W}{L}\right)(V_{GS} - V_T)^2 = \frac{1}{2}K_n(V_{DS} - V_T)^2. \qquad (3.37)$$

Due to the non-linear relation between the current flowing through the transistor, i_{DS}, and the voltage dropping on this transistor V_{DS}, we can say that this transistor represents a non-linear resistor. Such a resistor may be used as a load in the NMOS inverter

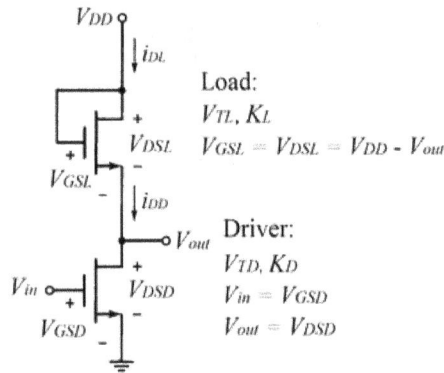

Fig. 3.30. An NMOS inverter with an enhanced mode load transistor.

circuit. Figure 3.30 represents an inverter based on NMOS technology using the enhanced mode transistor as a load.

The circuit presented in Fig. 3.30 has two transistors. The lower transistor is called the driver and the upper is called the load transistor. According to these definitions, all the parameters of the transistors have designations related to their functions, for example, V_{GSD} designates a gate-source voltage of the driver. Also, Fig. 3.30 presents the relation between individual transistor parameters and the parameters of the circuit: $V_{GSD} = V_{in}, V_{DSD} = V_{out}, V_{GSL} = V_{DSL} = V_{DD} - V_{out}$. This enables us to build inverter-circuit equations on the base of equations which describe the current states of the transistors.

Let us analyze the circuit presented in Fig. 3.30. The usual method of analysis is to apply two different extreme voltages to the input and calculate the output states. When the input voltage is zero, the transistor-driver will be in the cutoff state. Therefore, a current will not flow through both transistors, driver and load, $i_{DL} = i_{DD} = 0$. So, we can calculate the output voltage:

$$0 = i_{DSL} = \frac{1}{2}K_L(V_{GSl} - V_{TD})^2 = \frac{1}{2}K_L(V_{DSL} - V_{TL})^2$$

$$= \frac{K_L}{2}(V_{DD} - V_{out} - V_{TL})^2. \tag{3.38}$$

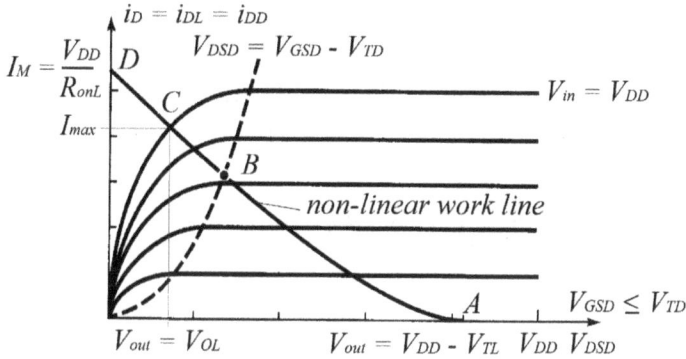

Fig. 3.31. The I–V characteristic of the NMOS inverter.

From this equation, we obtain

$$V_{out} = V_{DD} - V_{TL} = V_{OH} = \text{"1"}. \qquad (3.39)$$

Therefore, the maximum output voltage in our inverter cannot be more than $V_{DD} - V_{TL}$, which is called "poor 1". This point, A, should be designated in the load-line diagram.

If the input voltage is equal to $V_{in} = \text{"1"} = V_{DD}$, the transistor-driver will be in the state which enables the maximum current to be transferred. To estimate this current and analyze the circuit behavior, let us consider the I–V characteristic of our circuit which is presented in Fig. 3.31.

This figure represents a geometric solution to the joint behavior of two transistors, a load and a driver. The transistor-driver behavior is presented here in the form of its I–V characteristic family. A current flows through the circuit shown by the load-line AD. As the circuit is loaded with circuits of the same type, we may say that the currents flowing through both transistors are equal, $I_{DD} = I_{DL} = I_D$. As shown in the load-line diagram, the maximum virtual current, I_M, may be calculated if we assume that the internal resistance of the transistor-driver is zero, $R_{onD} = 0$. In this case, the supply voltage V_{DD} and the minimum internal resistance of the transistor-load R_{onL} will define point D in the following way using Eq. (3.23):

$$I_M = \frac{V_{DD}}{R_{onL}} = K_L V_{DD} (V_{DD} - V_{TL}). \qquad (3.40)$$

To find the coordinates of the point C, we should consider the states of both transistors. The transistor-driver is always in the triode state and the transistor-load is always in the saturation state. Therefore, one can write the following equation of currents:

$$i_{DD} = K_D \left[(V_{GSD} - V_{TD})V_{DSD} - \frac{1}{2}V_{DSD}^2 \right]$$

$$= \frac{1}{2}K_L \left(V_{GSL} - V_{TL} \right)^2 . \qquad (3.41)$$

For our circuit at point C when $V_{in} = V_{GSD} = V_{DD}$, and using the voltage definitions presented in Fig. 3.30, this equation may be rewritten as follows:

$$K_D[2(V_{DD} - V_{TD})V_{OL} - V_{OL}^2] = K_L(V_{DD} - V_{OL} - V_{TL})^2. \qquad (3.42)$$

This equation enables us to calculate the coordinates of point C with precision.

After calculating the extreme operating points, we can build a transfer function describing the behavior of the NMOS inverter shown in Fig. 3.30 on the base of its I–V characteristic. To achieve this goal, we will divide the transition performance during the switch from logic "1" to logic "0" into three regions defined by the transistors' behavior mode, as shown in Table 3.1.

In the first region, O-A, the function is constant and equal to $V_{OH} = V_{DD} - V_{TL}$ as shown in Eq. (3.39).

In the second region, A-B, we can obtain the function's behavior from the equity of the currents flowing through both transistors,

Table 3.1.   Transfer function behavior.

Region	V_{in}	Transistor-driver	Transistor-load
O-A	$V_{in} \leq V_{TD}$	Cutoff mode	Cutoff mode
A-B	$V_{in} \leq V_{out} + V_{TD}$	Saturation mode	Saturation mode
B-C	$V_{in} > V_{out} + V_{TD}$	Triode mode	Saturation mode

which are both in the saturation state:

$$i_{DD} = \frac{1}{2}K_D(V_{GSD} - V_{TD})^2 = \frac{1}{2}K_L(V_{GSL} - V_{TL})^2 = i_{DL}.$$

$$(3.43)$$

This equation, when rewritten for the inverter circuit, looks as follows:

$$K_D(V_{in} - V_{TD})^2 = K_L(V_{DD} - V_{out} - V_{TL})^2. \qquad (3.44)$$

The solution of this equation looks as follows:

$$V_{out} = V_{DD} - V_{TL} - \sqrt{\frac{K_D}{K_L}}(V_{in} - V_{TD}). \qquad (3.45)$$

Therefore, this function represents a straight line beginning from the coordinates $V_{in} = V_{TD}$ and $V_{out} = V_{DD} - V_{TL}$, $(V_{TD}, V_{DD} - V_{TL})$. This behavior is shown in Fig. 3.32.

A defined straight line continues up to point B when the transistor driver transits from the saturation mode to the triode mode. At this point $V_{DSD} = V_{GSD} - V_{TD}$ or in the circuit terminology, $V_{outB} = V_{inB} - V_{TD}$. The substitution of this condition into Eq. (3.45) enables

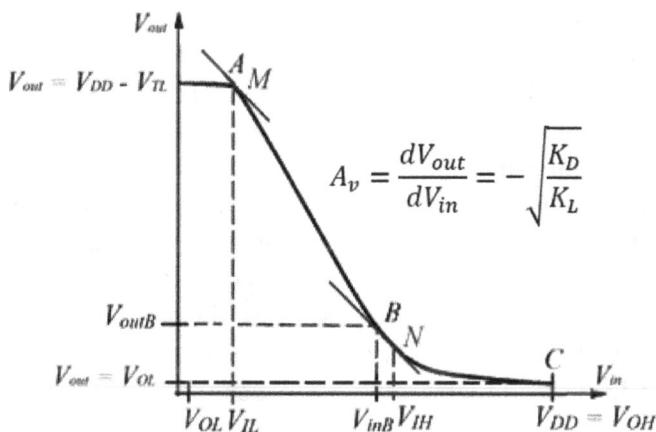

Fig. 3.32. The transfer function of the NMOS inverter with the enhanced mode load.

us to calculate the values of the point B coordinates:

$$V_{inB} = \frac{V_{DD} - V_{TL} + V_{TD}\left(1 + \sqrt{\frac{K_D}{K_L}}\right)}{1 + \sqrt{\frac{K_D}{K_L}}}. \qquad (3.46)$$

We obtain the equation describing the behavior of the transistors between points B and C from the relation $i_{DD} = i_{DL} = i_D$:

$$i_{DD} = K_D\left[(V_{GSD} - V_{TD})V_{DSD} - \frac{1}{2}V_{DSD}^2\right] = \frac{1}{2}K_L(V_{DSL} - V_{TL})^2 \qquad (3.47)$$

or, in the terms of the transfer function, this equation may be rewritten as follows:

$$K_D[2(V_{in} - V_{TD})V_{out} - V_{out}^2] = K_L(V_{DD} - V_{out} - V_{TL})^2. \qquad (3.48)$$

Evidently, the circuit behavior in this region is hyperbolic.

An analysis of the transfer function presented in Fig. 3.32 shows that the maximum voltage which remains in the circuit in the state of logic "1" is equal to $V_{IL} = V_{TD}$. Therefore, the points A and M coincide. The coordinates of the point N defining the minimum voltage remaining in the circuit in the low state, V_{IL}, may be calculated by differentiating Eq. (3.48). We calculate the noise margins using Eq. (1.4).

Now we can discuss the advantages and disadvantages of this inverter compared with the NMOS inverter with the resistive load. Evidently, this circuit has an important advantage due to the replacement of a load resistor with a transistor. A circuit without resistors is significantly smaller as the transistor takes up less space on the silicon wafer during the fabrication process. At the same time, this construction has several disadvantages. First of all, this is a "poor 1" i.e. low value of the V_{OH}. Second, the transfer function behaves linearly. In this case, we can define the magnification coefficient as follows using the assumption $L_D = L_L$ that is usually right in VLSI circuits fabricated by planar technology:

$$A_v = \frac{dV_{out}}{dV_{in}} = -\sqrt{\frac{K_D}{K_L}} = \sqrt{\frac{\left(\frac{W}{L}\right)_D}{\left(\frac{W}{L}\right)_L}} = \sqrt{\frac{W_D}{W_L}}. \qquad (3.49)$$

This shows that the transistor driver should have very large width relating to the transistor load to decrease the transition region between stable digital states. The third deficiency of this inverter is an asymmetric transfer function, which is very far from ideal. Therefore it makes sense to consider another type of inverter that uses a depletion type transistor as a load.

3.3.3. *The NMOS inverter with NMOS load of the depletion type*

Let us consider a depletion mode transistor NMOS with shortly connected gate and source, see Fig. 3.33. This connection is also called a diode-connected transistor.

An I–V characteristic of the diode-connected depletion mode NMOS transistor presented on Fig. 3.34 has been taken from the I–V characteristic family shown in Fig. 3.17. As shown, with the shortly connected gate and source, the transistor may take two different operation modes: a saturation mode and a triode mode. It is of interest that the voltage $V_{DSL} = 0$ if the current $i_{DL} = 0$.

To analyze the circuit behavior, assume that the input voltage, v_{in}, equals to zero. In this case, the transistor driver will be in the cutoff state and the current i_{DD} will be equal to zero also, $i_{DD} = 0$. A current through the load transistor $i_{DL} = i_{DD} = 0$, therefore

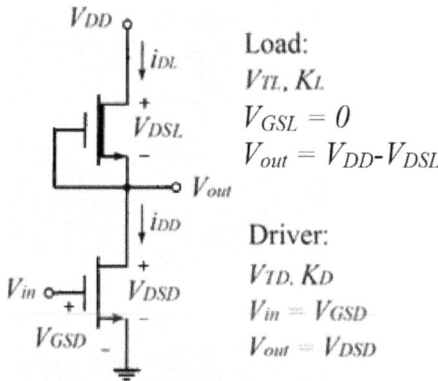

Load:
V_{TL}, K_L
$V_{GSL} = 0$
$V_{out} = V_{DD} - V_{DSL}$

Driver:
V_{TD}, K_D
$V_{in} = V_{GSD}$
$V_{out} = V_{DSD}$

Fig. 3.33. An NMOS inverter with a depletion-mode loading transistor.

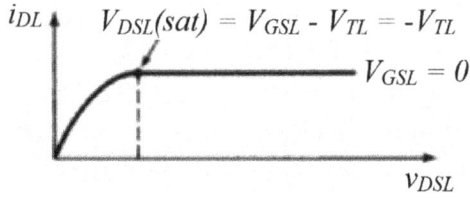

Fig. 3.34. An I–V characteristic of the diode-connected depletion mode NMOS transistor.

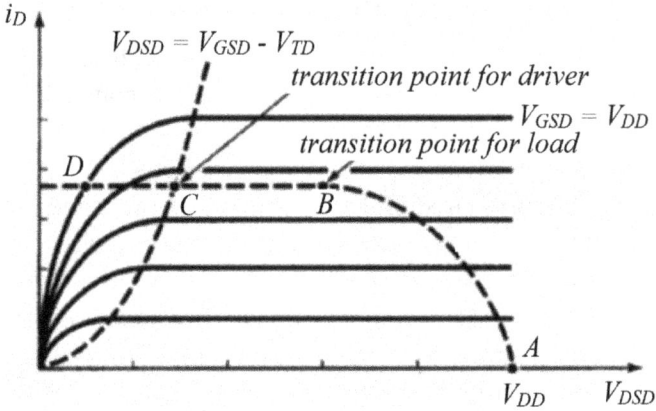

Fig. 3.35. An I–V characteristic of the NMOS inverter with the depletion-mode loading transistor.

$V_{DSL} = 0$ and $V_{out} = V_{DD} - V_{DSL} = V_{DD} = V_{OH} = $ "1". Therefore, the maximum output voltage enabled by the inverter is a "good 1". Figure 3.35 represents a superposition of the I–V characteristics of both transistors at the inverter circuit, Fig. 3.33.

The load-line of this inverter represents a part of the I–V characteristic of the loading transistor. This load line goes through several breaking points, $\boldsymbol{A \div D}$, where A and D are the transition points of both transistors. So, the line $\boldsymbol{A\text{-}B\text{-}C\text{-}D}$ is the geometric solution for the joint behavior of two transistors, a load and a driver. Here, point $\boldsymbol{A}$ designates the transition of the transistor driver from the cutoff state to the saturation mode. The transistor load is in the triode mode this time. At point $\boldsymbol{B}$, the transistor load enters the saturation mode and now both transistors are in the saturation mode. Point C designates the transition from the saturation mode to the triode mode for the driver transistor. Point D represents a boundary solution for the inverter. The maximum current flowing in this inverter will be constant at the line B-C-D.

We can build the transfer function of the inverter by using Fig. 3.35. If the input voltage V_{in} is low or equal to the value V_{TD}, the transistor-driver is in the cutoff state. In this region, the function is constant and equal to $V_{OH} = V_{DD}$. Here, the current in the circuit is zero. Point A designates the beginning of changes in the circuit behavior. For example, it is at this point that the driver enters the saturation mode.

In the second region, A-B, we can obtain the function behavior from the equity of the currents flowing through both transistors, when the driver is in the saturation mode and the load is in the triode mode:

$$i_{DD} = \frac{1}{2}K_D(V_{GSD} - V_{TD})^2 = \frac{1}{2}K_L[2(V_{GSL} - V_{TL})V_{DSL} - V_{DSL}^2].$$
$$(3.50)$$

This equation, rewritten for the inverter circuit, looks as follows:

$$K_D(V_{in} - V_{TD})^2 = K_L[2(-V_{TL})(V_{DD} - V_{out}) - (V_{DD} - V_{out})^2].$$
$$(3.51)$$

This non-linear behavior continues up to the point B, which designates the transition of the load from the triode mode to the saturation mode. From this point, both transistors are in the saturation mode and the following equation describes their behavior:

$$i_{DD} = \frac{1}{2}K_D(V_{GSD} - V_{TD})^2 = \frac{1}{2}K_L(V_{GSL} - V_{TL})^2 = i_{DL}. \quad (3.52)$$

This equation, rewritten for the inverter circuit, looks as follows:

$$K_D(V_{in} - V_{TD})^2 = K_L(-V_{TL})^2. \quad (3.53)$$

The solution of this equation looks as follows:

$$V_{in} = V_{TD} - \sqrt{\frac{K_L}{K_D}}(-V_{TL}). \quad (3.54)$$

We get great results. In the obtained equation, the input voltage is constant and the output voltage changes its value instantly. So,

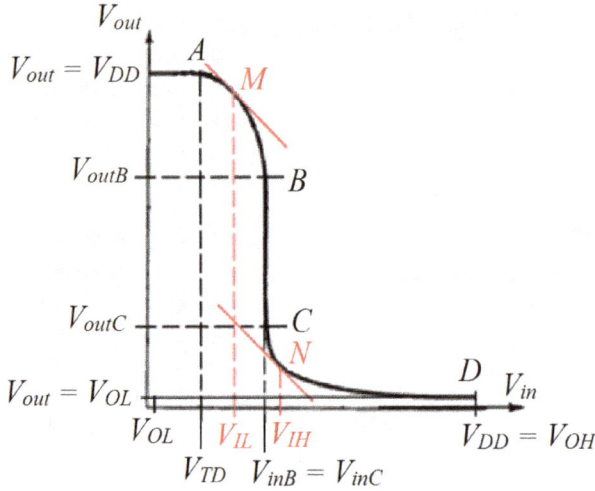

Fig. 3.36.   Transfer function of the NMOS inverter with the depleted type load.

$V_{inB} = V_{inC} =$ constant during the whole time that V_{out} changes from V_{outB} to V_{outC}. The transfer function of the circuit shown in Fig. 3.33 is presented in Fig. 3.36.

To calculate the output voltage at the points B and C, we need to use conditions defining these transition points and Fig. 3.35. At point B, the transistor load enters the saturation mode from the triode mode, therefore

$$V_{DSL} = V_{GSL} - V_{TL} = -V_{TL} = V_{DD} - V_{outB} \quad \text{or}$$

$$\boldsymbol{V_{outB} = V_{DD} + V_{TL}}. \tag{3.55}$$

It should be remembered that V_{TL} has a negative value. As for the output voltage at point C, when the driver changes the saturation mode to the triode mode, we can write as follows:

$$V_{outC} = V_{GSD} - V_{TD} = V_{inC} - V_{TD}$$

$$= V_{TD} - \sqrt{\frac{K_L}{K_D}}(-V_{TL}) - V_{TD}$$

$$= \sqrt{\frac{K_L}{K_D}}(-V_{TL}). \tag{3.56}$$

In the third region, C-D, the circuit behaves non-linearly again. For this region, when the driver is in the triode mode and the load is in the saturation region, we can equate the currents in the following way:

$$i_{DD} = \frac{1}{2}K_D[2(V_{GSD} - V_{TD})V_{DSD} - V_{DSD}^2] = \frac{1}{2}K_L(V_{DSL} - V_{TL})^2 \tag{3.57}$$

in terms of the transfer function, it will look as follows:

$$K_D[2(V_{in} - V_{TD})V_{out} - V_{out}^2] = K_L(V_{DD} - V_{out} - V_{TL})^2. \tag{3.58}$$

Evidently, the circuit behavior in this region is hyperbolic.

An analysis of the transfer function presented in Fig. 3.36 shows that the transfer between points B and C occurs instantly. The symmetry level in the circuit behavior is fully defined by the values of the threshold voltages V_{TD} and V_{TL} and by the transconductance of transistors K_D and K_L. Depending on the relation between parameters K_D and K_L, the position of the points B and C on the transfer characteristic curve may be changed as shown in Fig. 3.37.

The coordinates of the points M and N may be calculated by differentiating Eqs. (3.51) and (3.58). We calculate the noise margins

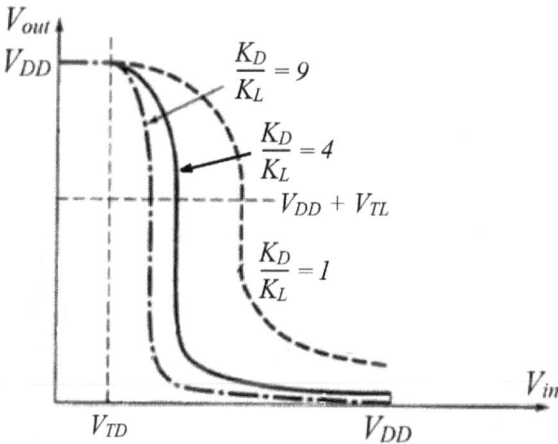

Fig. 3.37. Transfer characteristic dependence on the transconductance.

using Eq. (1.4). Evidently, this circuit has many advantages and it was applied in the industry.

3.4. Complementary MOS Technology (CMOS)

3.4.1. *The basic CMOS inverter, a principle of operation*

In the previous three parts of the chapter, we have considered three types of inverters based on the use of MOS transistors. In all these circuits, the transistor driver was of the NMOS enhanced mode type and only the load was different: a resistor, an enhanced mode NMOS, or a depletion mode NMOS. Figure 3.38 schematically presents the three technologies mentioned above. Each one has advantages and weaknesses however they are all far from being the ideal inverter.

Let us consider two transistors of different types, p and n, connected in series as shown in Fig. 3.39 and integrated together in the same silicon crystal. In this way, these transistors may be fabricated and connected through a common building process known as "planar" technology. This technology consists of a series of various processes enabling the growth of very complicated digital circuits on one surface of a silicon single-crystalline wafer.

As shown in Fig. 3.39, the source of the NMOS transistor is connected to the ground and the source of the PMOS transistor

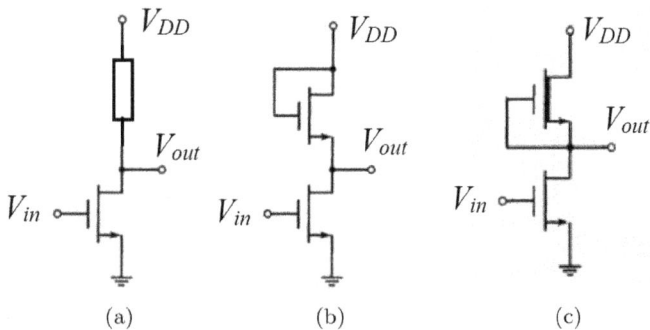

Fig. 3.38.   Three inverters based on NMOS technology.

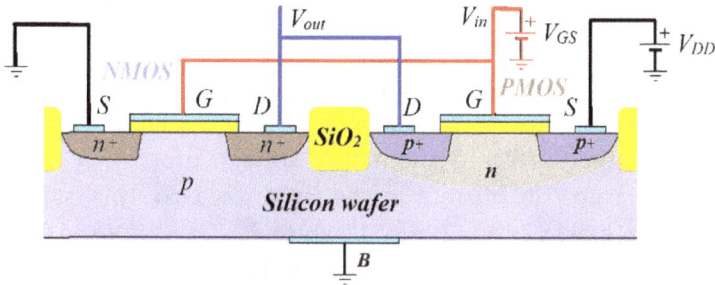

Fig. 3.39. Two transistors NMOS and PMOS connected in one circuit.

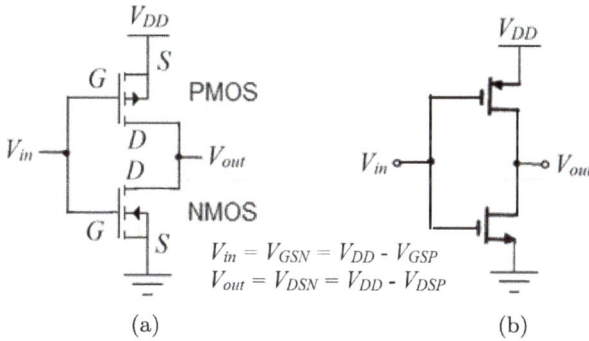

$$V_{in} = V_{GSN} = V_{DD} - V_{GSP}$$
$$V_{out} = V_{DSN} = V_{DD} - V_{DSP}$$

(a)   (b)

Fig. 3.40. Schematic representation of the NMOS-PMOS inverter.

is connected to the supply V_{DD}. The gates of both transistors are connected together and represent an input in the circuit. The drains of both transistors are also connected and this common contact represents an output from the circuit. Therefore, the construction shown in Fig. 3.39 represents an electronic circuit and it is shown in Fig. 3.40. Here, the NMOS transistor continues to work as a driver and the PMOS transistor plays the role of a load. Both transistors here are enhanced mode devices.

Let us consider the behavior of the circuit in Fig. 3.40. Firstly, assume that the input voltage is zero. In this case, the NMOS transistor will be in the cutoff state ($V_{GSN} < V_{TN}$) and a current through NMOS will be equal to zero. However, the full V_{DD} voltage will be applied to the input of the load transistor, PMOS: $V_{GSP} = V_{DD} - V_{in} = V_{DD}$. Therefore, the output voltage in this case will be equal to V_{DD}, as the current in the PMOS transistor will also be equal to zero, and the inversion layer of this transistor represents an equipotential surface without any voltage drop: $V_{out} = V_{OH} = V_{DD} = $ "1". Since the current in the circuit is zero, this circuit does not consume power in the logic "1" state.

If the input voltage is high, $V_{in} = V_{DD}$, the driver transistor NMOS will be in the triode state. However, when this voltage is applied to the load transistor PMOS, it will cause the load transistor PMOS to enter the cutoff state: $V_{GSP} = V_{DD} - V_{in} = V_{DD} - V_{DD} = 0 < |V_{TP}|$. Thus, the current in the circuit will be equal to zero again. The output voltage in this case will be zero: $V_{out} = V_{OL} = 0 = $ "0". The power consumed in the circuit is zero once more. Therefore, our circuit does not consume any power in both stable logic states.

We have seen that our circuit turns an input signal into the opposite, therefore this circuit behaves like an inverter. Now, let us build the load curve of the NMOS-PMOS inverter. To achieve this goal, we must present the I–V characteristics of both transistors together as shown in Fig. 3.41. The different colors in Fig. 3.41

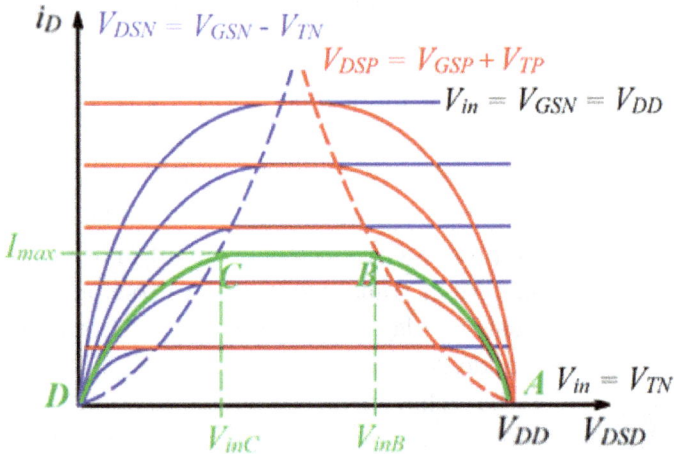

Fig. 3.41.   Current-voltage characteristics of the NMOS-PMOS inverter.

illustrate the characteristic families of both transistors: the blue curves represent the I–V characteristics of the NMOS transistor or the driver and the red curves represent the I–V characteristics of the load transistor, PMOS.

In this figure, point A represents a case in which the input voltage is zero. Here, the current through the circuit is zero also. The opposite state, when the input voltage $V_{in} = V_{GSD} = V_{DD}$, is illustrated by the point D. In this state, the current is also zero. As shown, the growth of the input voltage in the circuit leads to the transition of both transistors through various operation modes. We look at five different regions of the circuit's behavior:

- The first region is up to the point A, $V_{in} \leq V_{TN}$ (NMOS is cut off);
- The second region is A-B, $V_{TN} < V_{in} \leq V_{inB}$ (NMOS is saturated and PMOS in the triode mode);
- The third region is B-C, $V_{inB} \leq V_{in} \leq V_{inC}$ (both transistors are saturated);
- The fourth region is C-D, $V_{inC} \leq V_{in} < V_{DD} + V_{TP}$ (NMOS is in the triode mode and PMOS is saturated);
- The fifth region is from point D onward, $V_{DD} + V_{TP} \leq V_{in} \leq V_{DD}$ (PMOS is cut off).

Let us consider the properties of the transistors at the extreme points when $V_{in} = 0$ and $V_{in} = V_{DD}$. At the first point, $V_{in} = 0$, the NMOS transistor is in the cutoff state and the PMOS transistor is in the triode mode of operation, as shown in Fig. 3.41. So, we can write the equation describing the behavior of the transistor PMOS:

$$i_{DP} = \mu_p C_{ox} \left(\frac{W}{L}\right)_p \left[(V_{GSP} + V_{TP})V_{DSP} - \frac{1}{2}V_{DSP}^2\right]. \qquad (3.59)$$

As the value of the second summand in brackets is of the second degree of smallness, we can write the following approximation on condition that $V_{GSP} = V_{DD} - V_{in}$:

$$i_{DP} = \mu_p C_{ox} \left(\frac{W}{L}\right)_p [(V_{DD} + V_{TP})V_{DSP}]. \qquad (3.60)$$

From this equation we can find the minimum value of the resistance of the PMOS transistor:

$$R_{P,min} \equiv \frac{V_{DSP}}{V_{TP}} = \frac{1}{\mu_p C_{ox} \left(\frac{W}{L}\right)_p (V_{DD} + V_{TP})}$$

$$= \frac{1}{\mu_p C_{ox} \left(\frac{W}{L}\right)_p (V_{DD} - |V_{TP}|)}. \tag{3.61}$$

At the second point, $V_{in} = V_{DD}$, the PMOS transistor is in the cutoff state and the NMOS transistor is in the triode mode of operation, as shown in Fig. 3.41. So, we can write the equation describing the behavior of the transistor NMOS:

$$i_{DN} = \mu_n C_{ox} \left(\frac{W}{L}\right)_n \left[(V_{GSN} - V_{TN})V_{DSN} - \frac{1}{2}V_{DSN}^2\right]. \tag{3.62}$$

As the value of the second summand in brackets is of the second degree of smallness, we can write the following approximation on condition that $V_{in} = V_{GSN} = V_{DD}$:

$$i_{DN} = \mu_n C_{ox} \left(\frac{W}{L}\right)_n [(V_{DD} - V_{TN})V_{DSN}]. \tag{3.63}$$

From this equation we can find the minimum value of the resistance of the NMOS transistor:

$$R_{N,min} \equiv \frac{V_{DSN}}{V_{TN}} = \frac{1}{\mu_n C_{ox} \left(\frac{W}{L}\right)_n (V_{DD} - V_{TN})}. \tag{3.64}$$

The NMOS-PMOS inverter represents a capacitor if we consider it from the input point, as shown in Fig. 3.39. The gates of both transistors are connected together and separated from the transistors' bodies by a thin dielectric film such that we have capacity between the common gate and the ground. Therefore, circuits from the same family always load NMOS-PMOS inverters in the logic environment. Figure 3.42 represents this inverter loaded by the same inverter shown as a capacitor:

The inverter circuit presented in Fig. 3.42 may keep the logic state "0" or "1". In the high state, $V_{out} = $ "1", the output voltage is equal to V_{DD}. This voltage is on the capacitor C. The transition of the circuit from one logic state to another occurs through the discharge

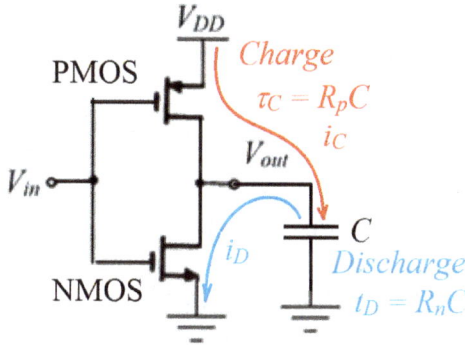

Fig. 3.42.   An inverter loaded by a capacitor.

or charge of the loading capacitor. To obtain the symmetrical type of the inverter's behavior, transition times for the charge and discharge should be equal, $\tau_D = \tau_C$ or $R_p = R_n$:

$$R_{N,min} = \frac{1}{\mu_n C_{ox} \left(\frac{W}{L}\right)_n (V_{DD} - V_{TN})}$$

$$= \frac{1}{\mu_p C_{ox} \left(\frac{W}{L}\right)_p (V_{DD} - |V_{TP}|)} = R_{P,min}. \qquad (3.65)$$

Eq. (3.65) includes several conditions which define the symmetrical behavior of the inverter:

1. The concentration of impurities in the parts p and n of the silicon wafer should be equal.
2. The material and thickness of the isolating layer in the gates of both transistors should be the same.
3. The threshold voltage of both transistors should be the same, $V_{TN} = |V_{TP}|$.
4. If we consider Eq. (3.65) once more, bearing in mind conditions 1–3 mentioned above, we obtain the following relation assuming that the length of both transistors is the same:

$$\mu_n w_n = \mu_p w_p. \qquad (3.66)$$

If both transistors satisfy conditions 1–4, they are called complementary MOS transistors or CMOS. The circuit of the inverter presented in Fig. 3.40 is that of a CMOS inverter. Sometimes, such an inverter is called a CMOS transistor.

3.4.2. *Transfer characteristics of the CMOS inverter*

Now, we can build the transfer function of the CMOS inverter using its I–V characteristic and the conditions which cause the complementary behavior of both transistors. As mentioned earlier, in the first region ($V_{in} \leq V_{TN}$), the NMOS transistor is in the cutoff state and the PMOS transistor is in the triode mode. Therefore, the output voltage here equals to the maximum voltage, $V_{out} = V_{OH} = V_{DD}$. The current in this region is zero.

After growing the input voltage until it is greater than the threshold voltage, the NMOS transistor will be in the saturation mode, but the PMOS transistor continues to be in the triode mode (**A-B** region). The following equation describes this behavior:

$$i_{DN} = \frac{1}{2}K_N(V_{GSD} - V_{TN})^2 = \frac{1}{2}K_P[2(V_{GSP} + V_{TP})V_{DSP} - V_{DSP}{}^2].$$

$$(3.67)$$

Here, K_N and K_P are the transconductance of the NMOS and PMOS transistors consequently.

This equation, when rewritten for the inverter circuit, looks as follows:

$$K_N(V_{in} - V_{TN})^2$$
$$= K_P[2(V_{DD} - V_{in} + V_{TP})(V_{DD} - V_{out}) - (V_{DD} - V_{out})^2].$$

$$(3.68)$$

This non-linear behavior continues up to the point B, which designates the transition of the PMOS transistor from the triode mode to the saturation mode. From this point, both transistors are in the saturation mode and the following equation describes the behavior of our inverter:

$$i_{DN} = \frac{1}{2}K_N(V_{GSN} - V_{TN})^2 = \frac{1}{2}K_P(V_{GSP} + V_{TP})^2 = i_{DP}.$$

$$(3.69)$$

This equation, when rewritten for the inverter circuit, looks as follows:

$$K_N(V_{in} - V_{TN})^2 = K_P(V_{DD} - V_{in} + V_{TP})^2. \qquad (3.70)$$

The solution of this equation looks as follows:

$$V_{in} = V_{inB} = V_{inC} = \frac{V_{DD} + V_{TP} + V_{TN}\sqrt{\frac{K_N}{K_P}}}{1 + \sqrt{\frac{K_N}{K_P}}}. \qquad (3.71)$$

We get this great result once more! The obtained equation shows that the input voltage is constant and the output voltage changes its value instantly in the region **B-C**. So, $V_{inB} = V_{inC}$ is constant all the time when V_{out} changes from V_{outB} to V_{outC}. If all four conditions of complementarity for NMOS and PMOS transistors are met, we obtain the CMOS inverter that the transfer function presents in Fig. 3.43.

In this case, $V_{TN} = |V_{TP}| = V_T$, which leads to the transition of Eq. (3.71) as follows:

$$V_{inB} = V_{inC} = 0.5V_{DD}. \qquad (3.72)$$

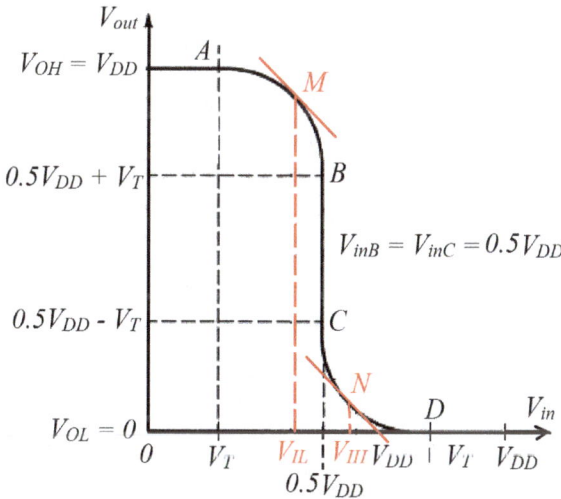

Fig. 3.43.   The transfer function of the CMOS inverter.

To calculate the output voltage related to the points $\boldsymbol{B}$ and $\boldsymbol{C}$, we must use equations that define the properties of these points. For the point $\boldsymbol{B}$, or the point of transition of the PMOS from the triode mode to the saturation mode, this equation is $V_{DSP} = V_{GSP} + V_{TP}$ or $V_{DD} - V_{outB} = V_{DD} - V_{inB} - V_T$ or $V_{outB} = V_{inB} + V_T$ and substituting in Eq. (3.72), $V_{outB} = 0.5V_{DD} + V_T$. In this way we can calculate the output voltage at the point $\boldsymbol{C}$. At this point, the transistor NMOS transits from the saturation mode to the triode mode. So, $V_{DSN} = V_{GSN} - V_{TN}$ or $V_{outC} = V_{inC} - V_T = 0.5V_{DD} - V_T$.

The CMOS inverter's transfer function is symmetrical. Therefore, after point $\boldsymbol{C}$ and up to the point $\boldsymbol{D}$, the inverter behaves in the same way as in the region $\boldsymbol{A}$-$\boldsymbol{B}$, but the transistors change places. Now, the NMOS transistor is in the triode mode and the PMOS transistor is in the saturation mode. An equation describing this behavior looks as follows:

$$i_{DN} = \frac{1}{2}K_N[2(V_{GSN} - V_{TN})V_{DSN} - V_{DSN}^2] = \frac{1}{2}K_P(V_{DSP} - V_{TP})^2.$$

$$(3.73)$$

Or in the terms of the transfer function, it will be hyperbolic:

$$K_N[2(V_{in} - V_T)V_{out} - V_{out}^2] = K_P(V_{DD} - V_{out} - V_T)^2. \quad (3.74)$$

After the point D, our inverter enters the low logical state when the transistor PMOS is closed and the current through the circuit is reduced to zero. Here, the output voltage is also equal to zero.

We will calculate the noise margins of the CMOS inverter in a known way. Let us calculate the input voltage V_{IL} for the point $\boldsymbol{M}$ when the derivative of the transfer function is equal to -1. Equation (3.68) describes the inverter behavior at this point. As the transistors are complements, their transconductance is equal, $K_N = K_P$. Thus, the derivative of this equation will be as follows:

$$(V_{in} - V_T) + (V_{DD} - V_{out})$$

$$= -(V_{DD} - V_{in} - V_T)\frac{dV_{out}}{dV_{in}} + (V_{DD} - V_{out})\frac{dV_{out}}{dV_{in}}. \quad (3.75)$$

The substitution of the relation $\frac{dV_{out}}{dV_{in}} = -1$ into this equation gives $V_{outM} = V_{IL} + 0.5V_{DD}$. Now, we substitute the value V_{outM} into

Eq. (3.68) and solve it for $V_{inM} = V_{IL}$:

$$V_{inM} = V_{IL} = \frac{1}{8}(3V_{DD} + 2V_T). \qquad (3.76)$$

V_{IH} for this inverter may be found using the symmetry of the transfer function:

$$V_{IH} - \frac{V_{DD}}{2} = \frac{V_{DD}}{2} - V_{IL}. \qquad (3.77)$$

From this equation, substituting in Eq. (3.76), we obtain:

$$V_{inN} = V_{IH} = \frac{1}{8}(5V_{DD} - 2V_T). \qquad (3.78)$$

From this relation we can calculate the noise margins:

$$NM_L = V_{IL} - V_{OL} = \frac{1}{8}(3V_{DD} + 2V_T). \qquad (3.79)$$

$$NM_H = V_{OH} - V_{IH} = \frac{1}{8}(3V_{DD} + 2V_T). \qquad (3.80)$$

As shown, these values are the maximum possible and equal, see Fig. 3.43.

Figure 3.44 represents the series of inverters based on the MOS transistors:

As can be seen, the main difference in these circuits is the loading device: a resistor, an NMOS transistor or a PMOS transistor. It is interesting to note that the last circuit, the CMOS inverter, has a transfer function, which is as close as possible to the ideal transfer function (see Fig. 1.13). The CMOS inverter has the following advantages in comparison with all the other inverters:

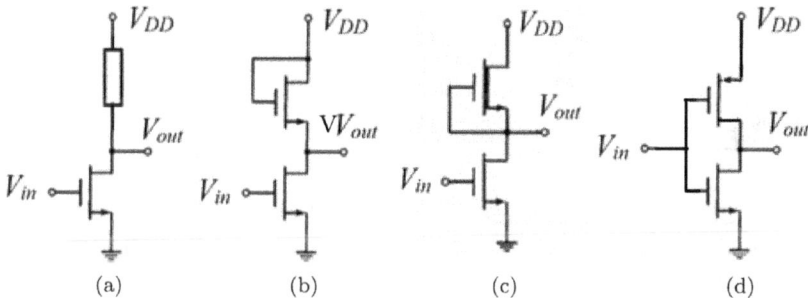

Fig. 3.44. Various types of MOS inverters.

- It does not consume energy in the static logical states;
- It has a symmetrical transfer function, so the transition time from one logic state into another is the same;
- It has the maximum possible noise margins for electronic inverters, and these noise margins are equal;
- It has the minimum dimensions and very well-developed technology and it integrates into the widely used silicon technology.

One of the important characteristics of inverters based on BJT transistors is a value of the maximum permissible fan-out. This value is limited by the maximum input resistance of the inverter. The input resistance of NMOS and CMOS inverters is very high, $\sim 10^{14}\Omega$, that enables us to connect an infinite set of logic gates to the output of such inverters without any influence on the logical state of the driving inverter. In these inverters, the limit of fan-out comes from another side. Due to the high input resistance, each of the loading logic gates may be seen as a capacitor with an average capacitance defined by the dielectric gate's layer dimensions and its dielectric properties. So, all loading circuits connected to the output contact will be represented by one common capacitor which includes all connected capacitors. In other words, the value of this common capacitor will be equal to the sum of parallel-connected capacitors as presented in Fig. 3.45.

As shown in Fig. 3.45, the operation time will limit the number of logic gates connected to the output of the inverter-driver based on CMOS transistors. Therefore, we can estimate a maximum fan-out for such circuits by using a defined time of operation. For example,

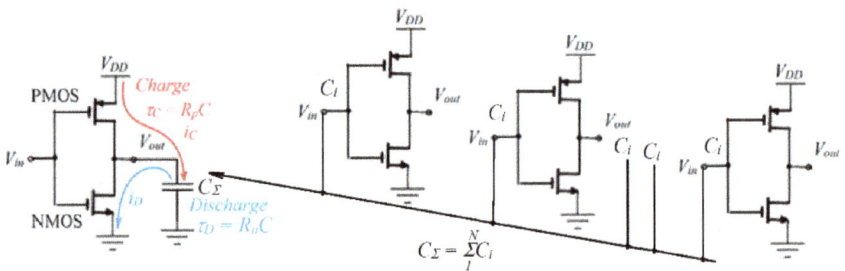

Fig. 3.45. A parallel connection of the load circuits to the inverter-driver.

let us assume that the input capacitance of each loading circuit is $C_i = 1fF = 10^{-15}F$, the minimum resistance of the MOS transistor providing charge or discharge of the load is equal to $R_n = R_p = 1$ kΩ and the operation time is 1 ns. We should estimate the number of loading circuits in the following way:

$$N_{max} = \frac{\tau}{RC_i} = \frac{10^{-9}}{10^3 10^{-15}} = 10^3 = 1000 \ circuits.$$

3.4.3.   The dynamic operation of the CMOS inverter

The propagation time and the time delay in the various logic circuits define the speed of operation of the computers and other digital systems. The inverter is the basic element of each digital technology. Therefore, the propagation time, t_p, is the fundamental parameter characterizing the digital system and the suitable digital technology. In the past, CMOS technology was the most common digital technology. Thus, we should estimate the propagation time delay in the CMOS inverter. To achieve this goal, assume that the output capacitor represents all possible loads for our inverter. Also, assume that an ideal rectangular pulse represents the input signal. Figure 3.46(a) represents the CMOS inverter loaded by the

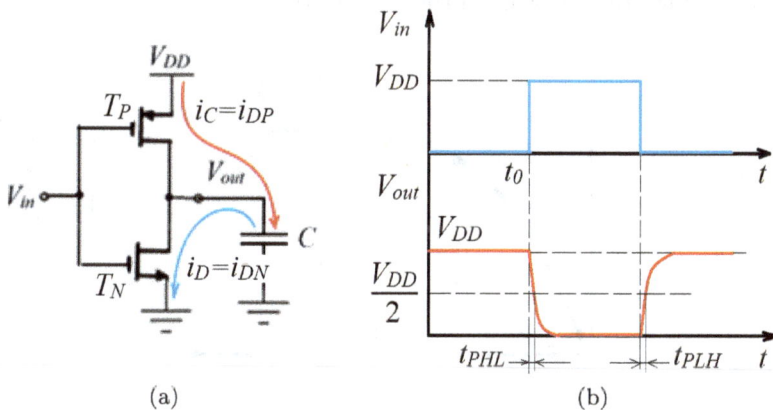

Fig. 3.46.   The dynamical behavior of the CMOS inverter: (a) an inverter loaded by the capacitor, (b) both the input (blue) and output (red) signals.

capacitor C and Fig. 3.46(b) represents the behavior of the input pulse and output pulse passing through the inverter.

By definition (see Fig. 1.11), we estimate the time propagation delays t_{PHL} and t_{PLH} for the decreasing and increasing pulses to range from half of the maximum value of the input pulse up to half of the maximum value of the output pulse. As a CMOS inverter is a symmetric circuit, it is enough for us to estimate only one of two parameters, t_{PHL} or t_{PLH}. Let us consider the input and output pulses shown in Fig. 3.46(b). As shown, the inverter presented in Fig. 3.46(a) is in the steady state with $V_{in} = 0$ and $V_{out} = V_{DD}$ up to time t_0. Exactly at the time t_0, the input voltage immediately rises up to $V_{in} = V_{DD}$. The output voltage at once begins to decrease. This behavior is illustrated in Fig. 3.47. Here, we can see the changes to the current when the output voltage is decreased. At the time t_0, the input voltage $V_{in} = V_{GSN}$ jumps from 0 up to V_{DD}. This is illustrated by the transition of the working point of the circuit from point A to point B. Transistor NMOS enters in the saturation mode and a load capacitor C begins to discharge through the NMOS transistor. His discharge occurs in two stages. In the first stage, the current is constant due to the saturation mode of the transistor's work. This behavior is illustrated by changes within section BC of the I–V curve, see Fig. 3.47. Point C corresponds with the transition of the NMOS transistor from the saturation mode to the triode mode

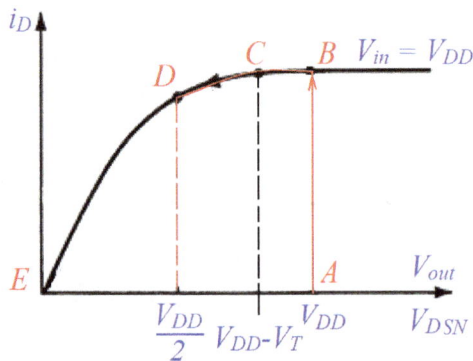

Fig. 3.47.   The current behavior in the CMOS inverter through transition from "1" to "0".

of operation, $V_{DSN} = V_{GSN} - V_T$. In the second stage, section CD, the current discharges are non-linear. Point D corresponds with the output voltage equal to half of the maximum, $V_{out} = V_{DSN} = 0.5 V_{DD}$. Therefore, the full time taken by the transition will be equal to the combined time taken by both stages, $t_{PHL} = t_{PHL1} + t_{PHL2}$.

We can find the time taken for the first stage of discharge (section BC) from the general relation $Q = it = CV$, taking into account the constant current of discharge in the saturation mode of operation of the transistor NMOS:

$$t_{PHL1} = \frac{CV}{i} = \frac{C(V_B - V_C)}{i_{DN}} = \frac{C[V_{DD} - (V_{DD} - V_T)]}{\frac{1}{2}K'_n \left(\frac{W}{L}\right)_n (V_{GSN} - V_T)^2}$$

$$= \frac{CV_T}{\frac{1}{2}K'_n \left(\frac{W}{L}\right)_n (V_{DD} - V_T)^2}. \tag{3.81}$$

In the second stage, the discharge current behaves according to the following equation:

$$i_{DN} = -C\frac{dV_{out}}{dt} \tag{3.82}$$

which may be rewritten as follows:

$$K'_n \left(\frac{W}{L}\right)_n \left[(V_{GSN} - V_T)V_{DSN} - \frac{1}{2}V_{DSN}^2\right] = -C\frac{dV_{DSN}}{dt} \tag{3.84}$$

or

$$K'_n \left(\frac{W}{L}\right)_n \left[(V_{DD} - V_T)V_{out} - \frac{1}{2}V_{out}^2\right] = -C\frac{dV_{out}}{dt}. \tag{3.85}$$

Now, we can separate the variables:

$$-\frac{K'_n \left(\frac{W}{L}\right)_n}{2C}dt = \frac{dV_{out}}{2(V_{DD} - V_T)V_{out} - V_{out}^2} \tag{3.86}$$

and rewrite this equation as follows:

$$-\frac{K'_n \left(\frac{W}{L}\right)_n}{2C}dt = \frac{1}{2(V_{DD} - V_T)}\frac{dV_{out}}{\frac{1}{2(V_{DD}-V_T)}V_{out}^2 - V_{out}}. \tag{3.87}$$

The obtained form of the right side of Eq. (3.87) corresponds with the well-known tabular integral $\int \frac{dx}{ax^2 - x} = ln(1 - \frac{1}{ax})$. Therefore, we can write the solution of Eq. (3.87) as follows:

$$\frac{K_n' \left(\frac{W}{L}\right)_n}{2C} t_{PHL2} = \frac{1}{2} ln \left(\frac{3V_{DD} - 4V_T}{V_{DD}}\right) \quad \text{or}$$

$$t_{PHL2} = \frac{c}{K_n' \left(\frac{W}{L}\right)_n (V_{DD} - V_T)} ln \left(\frac{3V_{DD} - 4V_T}{V_{DD}}\right).$$

$$(3.88)$$

The full propagation delay will be equal to $t_{PHL} = t_{PHL1} + t_{PHL2}$ or substituting this equation into Eqs. (3.81) and (3.88), we obtain the following:

$$t_{PHL} = \frac{c}{K_n' \left(\frac{W}{L}\right)_n (V_{DD} - V_T)} \left[\frac{V_T}{V_{DD} - V_T} + \frac{1}{2} ln \left(\frac{3V_{DD} - 4V_T}{V_{DD}}\right)\right].$$

$$(3.89)$$

If we consider that usually the threshold voltage V_T for MOS transistors is approximately 20% of the supply voltage V_{DD}, i.e. $V_T = 0.2 \, V_{DD}$, substituting this value into Eq. (3.89) leads to the simplified reduced equation

$$t_{PHL} = \frac{1.6C}{K_n' \left(\frac{W}{L}\right)_n V_{DD}}.$$

$$(3.90)$$

The obtained Eq. (3.90) is only right for the pulse transition through the CMOS inverter.

To evaluate the propagation delay for the CMOS inverter, let us take the following numerical values for the parameters included in the t_{PHL}: $K_n = 20 \, \mu A/V^2$, $C = 20$ pF, $V_{DD} = 5$ V, $V_T = 1$ V. The substitution of these parameters into Eq. (3.90) shows that $t_{PHL} \approx 0.32 \, \mu s$. Thus, the propagation delay in the modern devices reaches several nanoseconds.

A result similar to that in Eq. (3.90) may be obtained in another way, through approximation. It is based on the simple averaging of the current flowing through the NMOS transistor while the capacitor discharges. Let us consider Fig. 3.47 once more. At the beginning point, A, the NMOS transistor is in the saturation mode, described by Eq. (3.19). At the finishing point, D, the NMOS transistor is in the triode mode, described by Eq. (3.20). Therefore, we can estimate the average current as the sum of the currents mentioned. If we take into account the relation between a supply voltage and a threshold voltage $V_T = 0.2 \ V_{DD}$, this estimation will look as follows:

$$
\begin{aligned}
\check{I}_{DN} &= \frac{1}{2}[I_{DN}(0) + I_{DN}(t_{PHL})] \\
&= \frac{1}{2} \left\{ K'_n \left(\frac{W}{L} \right)_n \left[(V_{GSN} - V_T)V_{DSN} - \frac{1}{2}V_{DSN}^2 \right] \right. \\
&\left. + \frac{1}{2} K'_n \left(\frac{W}{L} \right)_n (V_{GSN} - V_T)^2 \right\} \approx \frac{1}{4} K'_n \left(\frac{W}{L} \right)_n V_{DD}^2 .
\end{aligned}
$$
(3.91)

According to the general relation $Q = it = CV$, we can calculate the propagating time using the average current (see Eq. (3.91)) as follows:

$$
t_{PHL} = \frac{C\Delta V}{\check{I}_{DN}} = \frac{C\frac{V_{DD}}{2}}{\frac{1}{4}K'_n \left(\frac{W}{L} \right)_n V_{DD}^2} \approx \frac{1.7C}{K'_n \left(\frac{W}{L} \right)_n V_{DD}} .
$$
(3.92)

The result of the simplified estimation, Eq. (3.92), is similar to the result of the more precision calculation, Eq. (3.90) This enables us to use both methods for the calculation of the propagation delay.

Power consumption is one of the critical parameters of electronic circuits. We know that the CMOS digital circuits do not consume any power in the stable logic states. Thus, the currents flow in these circuits when the circuits transition from one logic state to another. Therefore, this power consumption during the transition of logic gates is also the dynamical behavior of the CMOS circuits. To estimate the power consumption by this dynamical behavior, let us consider the circuit presented in Fig. 3.48.

The transition of the inverter from the low state to the high state consists of the charging of the loading capacitor. The charge accumulated by the capacitor goes towards the polarization of a dielectric material in the capacitor. In other words, the charge energy goes towards the creation of dipoles in the dielectric material filling the capacitor. Figure 3.49 illustrates the charging process.

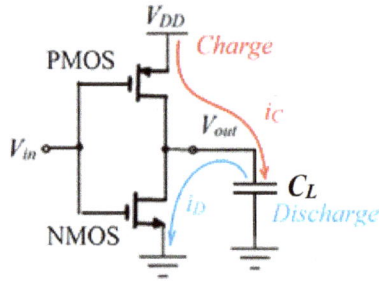

Fig. 3.48.   A CMOS inverter with a load.

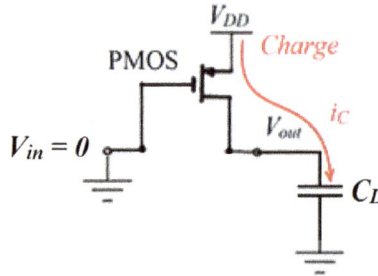

Fig. 3.49.   The charging of the load capacitor.

In this case, the non-linear current i_C flows through the PMOS transistor to charge the capacitor C_L:

$$i_C = C_L \frac{dV_{out}}{dt}. \tag{3.93}$$

The power P_p dissipated in the PMOS transistor through this process is equal to:

$$P_p = i_C V_{SD} = i_C(V_{DD} - V_{out}) = C_L \frac{dV_{out}}{dt}(V_{DD} - V_{out}). \tag{3.94}$$

Using this equation, one can estimate the energy on the capacitor as follows:

$$E_p = \int_0^\infty P_p dt = \int_0^\infty C_L(V_{DD} - V_{out})\frac{dV_{out}}{dt} dt$$

$$= C_L V_{DD} \int_0^{V_{DD}} dV_{out} - C_L \int_0^{V_{DD}} V_{out} dV_{out}$$

$$= \frac{1}{2} C_L V_{DD}^2. \tag{3.95}$$

The discharging process of the capacitor is illustrated in Fig. 3.50. Considering this figure and calculating the energy dissipated in the transistor NMOS in a similar way as before, we find that it is equal to the following:

$$E_n = \frac{1}{2} C_L V_{DD}^2. \tag{3.96}$$

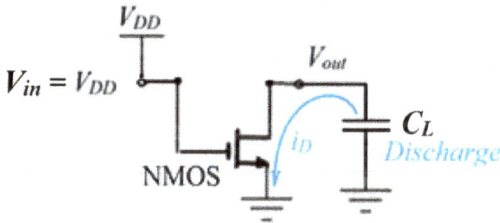

Fig. 3.50.   The discharging of the load capacitor.

Thus, the full energy dissipated through one cycle of charging-discharging will be the sum of E_p and E_n:

$$E_T = E_p + E_n = C_L V_{DD}^2. \tag{3.97}$$

If the process of charging-discharging occurs with the frequency f, we can calculate the power consumed by the CMOS circuit as follows:

$$P = f E_T = f C_L V_{DD}^2. \tag{3.98}$$

The maximum current flowing through the transition process may be calculated using Fig. 3.41 (an I–V characteristic of the CMOS inverter), Fig. 3.43 (a transfer function of the CMOS inverter), and Eq. (3.69), describing the behavior of the CMOS inverter when both transistors are in the saturation mode. As shown in the I–V characteristic, a maximum current is constant and maximum when both transistors, PMOS and CMOS, are in the saturation mode.

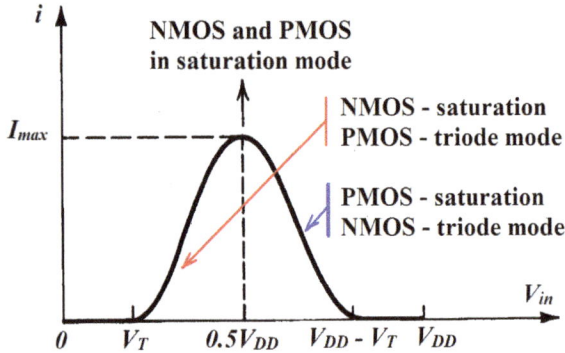

Fig. 3.51.   Input diagram for the CMOS inverter transition.

Therefore, from Eq. (3.69) we can calculate this current in the conditions described in the transfer function. Both transistors are in the saturation mode when $V_{in} = 0.5V_{DD}$, so the current for the CMOS inverter will be calculated as follows:

$$i_{DN} = \frac{1}{2}K_N(V_{GSN} - V_{TN})^2; \quad I_{max} = \frac{1}{2}K(0.5V_{DD} - V_T)^2.$$
$$(3.99)$$

To evaluate the CMOS inverter current behavior through the transition process, we need to consider Eqs. (3.68) and (3.74). These equations describe a current flowing through the circuit under an input voltage growth from $V_{in} = 0$ up to V_{DD}. Figure 3.51 illustrates this behavior.

As shown, a curve describing the current in the range of input voltage from $V_{in} = V_T$ up to $V_{in} = 0.5V_{DD}$, is defined by Eq. (3.68) when the NMOS transistor is in the saturation mode and the PMOS transistor is in the triode mode. Equation (3.74) defines a current flowing through the circuit with a growth of input voltage from $V_{in} = 0.5V_{DD}$ up to $V_{in} = V_{DD} - V_T$.

3.5.   Problems

3.5.1 Figure 3.52 presents a logic circuit with the following parameters: $V_{TNQ2} = V_{TNQ3} = 0.8$ V, $V_{TNQ1} = -2$ V, $(W/L)_{Q2} = (W/L)_{Q3} = 4$, $(W/L)_{Q1} = 1$, $k'_n = \mu_n C_{ox} = 35$ $\mu A/V^2$.

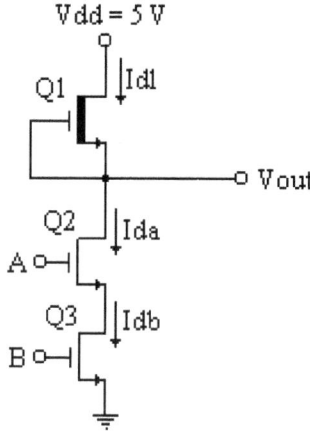

Fig. 3.52.   Logic circuit.

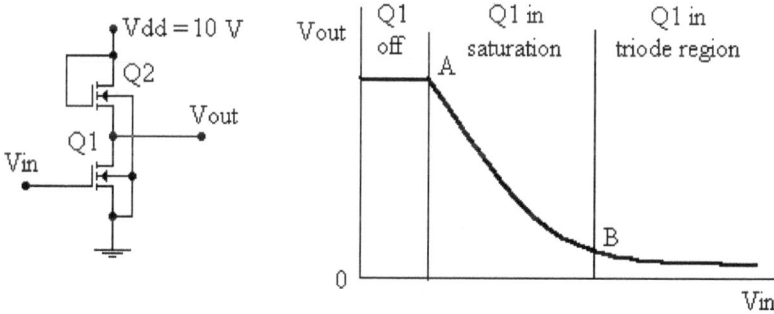

Fig. 3.53.   Logic circuit and the corresponding transfer function.

a. Identify the logic function implemented by the presented circuit.
b. Calculate the output voltage for two cases: (a) A = "0", B = "1"; (b) A = B = "1".

3.5.2 Figure 3.53 presents a logic circuit with the following parameters: $L_1 = 6$ μm, $L_2 = 30$ μm, $W_1 = 100$ μm, $W_2 = 6$ μm, $V_{T1} = V_{T2} = 1$V.

a. Calculate the coordinates of the points A and B.
b. Calculate the surface area of the circuit.

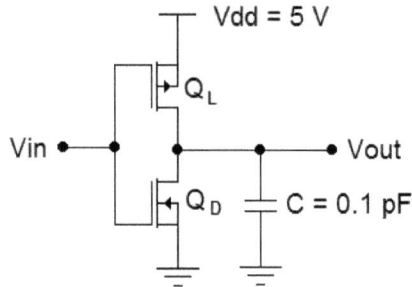

Fig. 3.54. Inverter logic circuit.

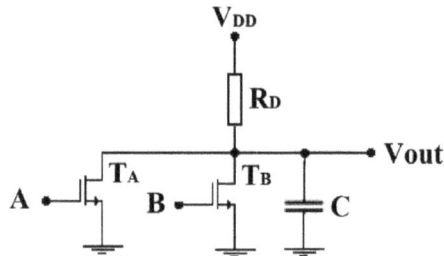

Fig. 3.55. Logic circuit with the capacitance load.

3.5.3 Figure 3.54 presents a logic circuit with the following parameters: $(W/L)n = 4/2$, $V_{tn} = -V_{tp} = 0.8$ V, $\mu_n Cox = 3\,\mu_p Cox = 50\,\mu A/V^2$, $(W/L)p = 10/2$.

    a. Calculate the noise margins of this circuit.

    b. Calculate a current flowing through the circuit for the following input voltages: $V_{in} = 0$, 2.5V, 5 V.

    c. Calculate the propagation time t_{PHL} for this circuit.

3.5.4 Figure 3.55 presents a logic circuit with the following parameters: $(W/L)n = 2$, $K'_n = 50\,\mu A/V^2$, $V_{Tn} = 1V$, $V_{DD} = 5$ V, $R_D = 50\,k\Omega$, $C = 5$ pF.

    a. Calculate the minimal output voltage.

    b. Calculate the propagation times t_{PHL} and t_{PLH}.

3.5.5 The parameters of the transistors Q1 and Q2 presented in Fig. 3.56 are as follows: $L1 = L2 = 10\,\mu m$, $\mu_n Cox = 20\,\mu A/V^2$, $V_T = 2$ V.

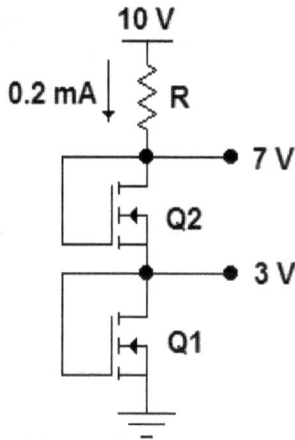

Fig. 3.56. Logic circuit.

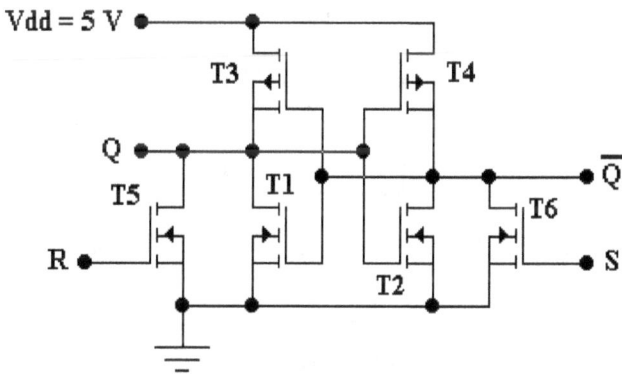

Fig. 3.57. RS-latch circuit.

Calculate the width of both transistors and the value of the resistor R.

3.5.6 Figure 3.57 presents a latch circuit with the following parameters: $|V_t| = 1$ V, K1 = K2 = K3 = K4 = K.

Calculate K5 = K6 if the circuit changes its state with input signals R or S equal to $V_{DD}/2$.

3.5.7 Figure 3.58 presents a logic circuit with the following parameters: $V_{TNQ6} = V_{TNQ7} = -1$, $V_{TNQ1} = V_{TNQ2} = V_{TNQ3}$

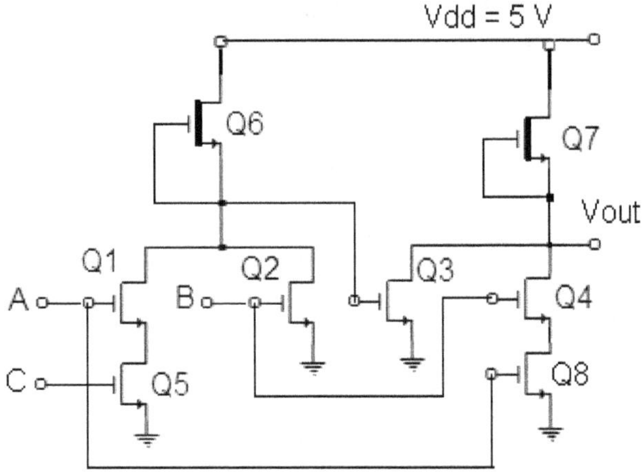

Fig. 3.58.   Logic circuit.

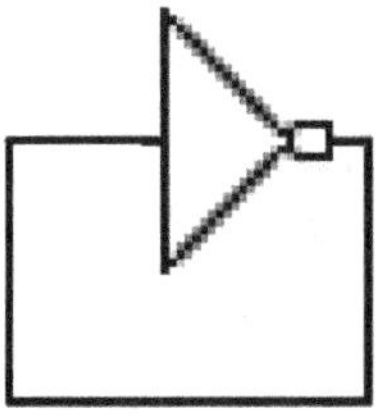

Fig. 3.59.   CMOS inverter with connected input and output.

$= V_{TNQ4} = V_{TNQ5} = V_{TNQ8} = 1$ V, $(W/L)_{Q6} = (W/L)_{Q7} = 3$,
$k'_n = \mu_n C_{ox} = 20$ $\mu A/V^2$, $(W/L)_{Qi} = 2$.

a. Identify the logic function implemented by the presented circuit.

b. Calculate the minimal output village.

3.5.8 It is known that in the CMOS inverter presented in Fig. 3.59, $V_{dd} = 5$ V, $V_t = 0.2$ V_{dd}, $\mu_n C_{ox} = 25$ $\mu A/V^2$, $(W/L) = 1.6$. With the help of the transfer function and the I–V characteristics, define the state of the circuit and calculate the current flowing through the circuit.

Chapter 4

Analysis and Synthesis of Digital Logic Circuits

4.1. Combinational circuits

4.1.1. *Basic principles of logic design in CMOS technology*

Figure 4.1 presents a basic scheme of the combinational logic circuit. The main property of combinational logic circuits is the dependence of each of the outputs on the input signal values. Thus, each change in the value of the input signals immediately changes the state of the output signals. There are many examples of combinational circuits: multiplexers, encoders and decoders, adders and multipliers, comparators, etc.

At the same time, we know that each logic function can have two different values only, "1" or "0". Therefore, if we take the combinational circuit with only one output, for the sake of simplicity, this circuit may be presented such as is shown in Fig. 4.2.

Here, we present a combinational circuit as a combination of two parts: a pull-up network (PUN), which comprises all the PMOS transistors applied in the circuit and a pull-down network (PDN), which comprises all the NMOS transistors applied in the circuit. In other words, if we assume that a supply is always at the top of the circuit and the ground is always at the bottom, a PUN comprises all the elements of the circuit which are between the supply and the

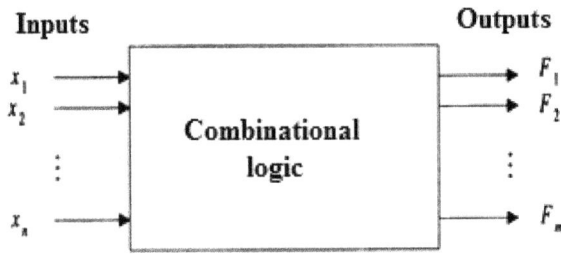

Fig. 4.1.   The general definition of the combinational logic circuit.

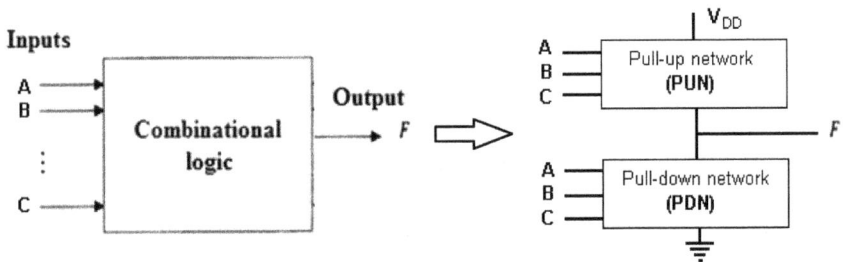

Fig. 4.2.   The general definition of the CMOS combinational logic circuit.

output line. A PDN comprises all elements of the circuit which are between the output line and the ground. By this logic, each one of the networks implements a suitable function: a PUN implements the function F and a PDN implements the complementary function, $\bar{F}$. In this construction, if all input variables $(A, B, \ldots C)$ are in the high state or "1", all the NMOS transistors which are in the pull-down network will be in the conductive (triode) mode of operation and all the PMOS transistors which are in the pull-up network will be in the cutoff state. Thus, the output point, F, will be isolated from the supply and connected with the ground. In other words, the output function will be equal to zero. If all input variables $(A, B, \ldots C)$ are in the low state or "0", all the NMOS transistors will be in the cutoff state and all the PMOS transistors will be in the conductive (triode) mode of operation. Therefore, the output point, F, will be connected with the supply and isolated from the ground. So, the output function will be in the high state, "1". Therefore, in

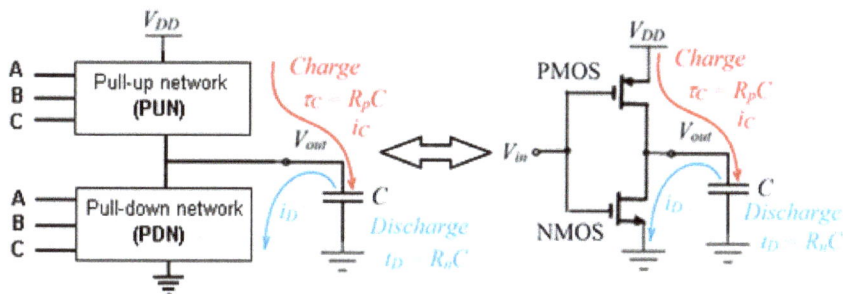

Fig. 4.3. Presentation of a combinational circuit as a CMOS inverter.

general, such a combinational circuit will work as an inverter and we can illustrate the work of the circuit as shown in Fig. 4.3.

The main idea behind a CMOS logic design is to build the pull-down circuit implementing a function inverse to the required one and then to build a PUN as the complement to the PDN using De Morgan's theorem:

$$\overline{(A + B)} = \bar{A} \cdot \bar{B}; \ \overline{A \cdot B} = \bar{A} + \bar{B}. \tag{4.1}$$

Using this conversion, we will obtain a logic circuit built from pull-up and pull-down networks which behaves as a simple CMOS inverter with equal charge and discharge times. Therefore, the transfer function of the built circuit will be symmetric and close to ideal. A simple CMOS inverter should be connected to the output of the built circuit to return the inverse function to the required one.

Let us consider the building of various logic circuits using the method mentioned above. Let us begin with the function $F = AB$. As defined, we built the function inverse to the required one, that is $F = \overline{AB}$, in the PDN. We obtain a complement function using De Morgan's theorem: $F = \bar{A} + \bar{B}$. As multiplication designates the serial connection of the transistors and sum designates the parallel connection of the transistors, the obtained circuit will look as shown in Fig. 4.4. A CMOS inverter connected to the output returns the circuit to the required function. This is illustrated using a truth table of the function. The PMOS transistors in the PUN play the role of complemented variables.

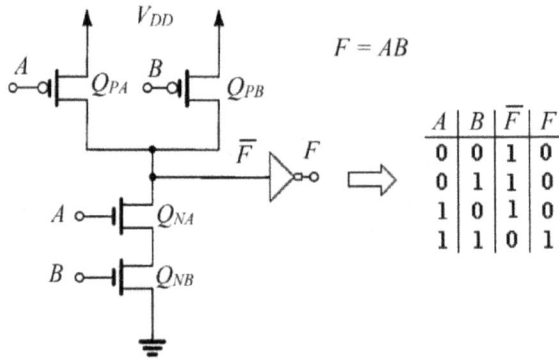

$F = AB$

A	B	$\overline{F}$	F
0	0	1	0
0	1	1	0
1	0	1	0
1	1	0	1

Fig. 4.4.   A function AND designed with CMOS technology.

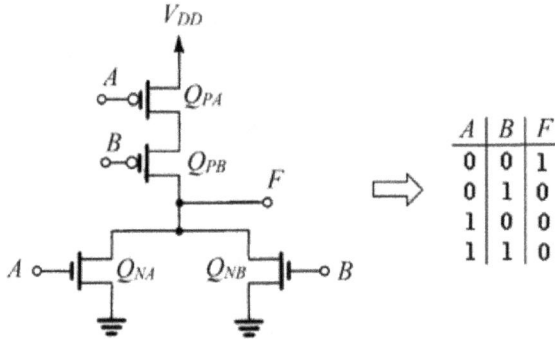

A	B	F
0	0	1
0	1	0
1	0	0
1	1	0

Fig. 4.5.   A function NOR designed with CMOS technology.

Sometimes, the required function is an inversion of the same logic function, for example $F = \overline{A + B} = \bar{A} \cdot \bar{B}$. In such a case, there is no reason to attach the inverter to the circuit output. This function, a logical sum, is presented in Fig. 4.5.

If we have to build more complex functions, for example $F = A + BC$, we build an inversion function in the PDN and a complementary function, through De Morgan's theorem, which is in the PUN as follows: $F = \overline{A + BC}(PDN) = \bar{A}(\bar{B} + \bar{C})(PUN)$. Figure 4.6 presents the circuit built according to this equation.

If the function which should be designed consists of both types of variable parameters, direct and complement, we must add the input

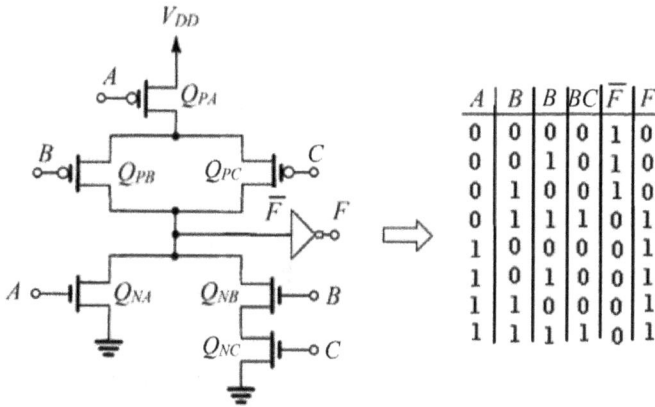

Fig. 4.6.   A complex function designed with CMOS technology.

A	B	B	BC	$\bar{F}$	F
0	0	0	0	1	0
0	0	1	0	1	0
0	1	0	0	1	0
0	1	1	1	0	1
1	0	0	0	0	1
1	0	1	0	0	1
1	1	0	0	0	1
1	1	1	1	0	1

inverters describing these complement variables. Let us consider the logical sum with modulus 2 (exclusive OR or XOR):

$$F = A \oplus B = \bar{A}B + A\bar{B}; \quad \bar{F} = AB + \bar{A}\bar{B} = A \otimes B. \qquad (4.2)$$

The input variables of this function are complement functions; therefore, we need to attach two invertors to the circuit for these variable parameters. Figure 4.7 presents the circuit built using CMOS technology. As shown in the figure, two additional inverters provide complementary input signals. It is of interest to calculate the number of transistors included in the circuit. We see here a total of 12 transistors: 6 PMOS and 6 NMOS transistors, where both input inverters consist of NMOS-PMOS pairs. This design method is simple and enables us to build electrical circuits realizing logical functions of any complexity. Moreover, this method enables the designer to calculate the area of the semiconductor wafer occupied by the designed electrical circuits.

4.1.2.   *Practical calculation of the area occupied by the CMOS circuit*

The main idea of the design of any logical circuits using CMOS technology is to build a circuit with the symmetric transfer function

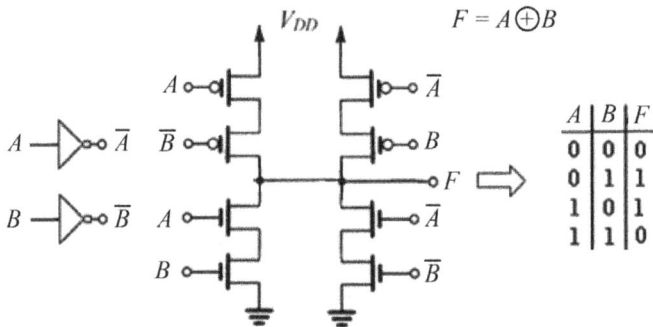

Fig. 4.7.   Exclusive OR function designed with CMOS technology.

repeating the transfer function of the basic or equivalent CMOS inverter. The main requirement of the PUN and PDN is their resistance equality. Thus, $R_{PUN} = R_{PDN}$. That provides the equal charge and discharge times of the capacitive load as shown in Fig. 4.3. At the same time, to find the area required for the circuit, we are obliged to equate the charge-discharge times for the designing circuit and the equivalent CMOS inverter. Therefore, we can describe this requirement in the following form:

$$R_{NMOSe} = R_{PDN} = R_{PUN} = R_{PMOSe}. \tag{4.3}$$

Here, the index e designates that the CMOS inverter is equivalent to our circuit. In other words, the charging and discharging times for our circuit are the same for the equivalent inverter, or the transfer functions of both circuits are the same. According to Eq. (3.23), the resistance of the open transistor is described by the formulae:

$$R_{NMOS} = \frac{1}{K_n(V_{DD} - V_T)} = \frac{1}{K'_n \left(\frac{w}{L}\right)_n (V_{DD} - V_T)} \quad \text{and}$$

$$R_{PMOS} = \frac{1}{K_p(V_{DD} - V_T)} = \frac{1}{K'_p \left(\frac{w}{L}\right)_p (V_{DD} - V_T)}. \tag{4.4}$$

Now, let us calculate the resistances of the built logic circuit parts, R_{PDN} and R_{PUN}. As is known, each electrical circuit is a combination of the series and parallel connections of various elements. Our elements are transistors, the NMOS transistors in the

PDN and the PMOS transistors in the PUN. Let us assume that we have two NMOS transistors connected in series as shown in Fig. 4.4. In this case, the full resistance of the PDN will be as follows:

$$R_{PDN} = R_{nA} + R_{nB} = \frac{1}{K'_n \left(\frac{w}{L}\right)_{nA} (V_{DD} - V_T)}$$

$$+ \frac{1}{K'_n \left(\frac{w}{L}\right)_{nB} (V_{DD} - V_T)}. \tag{4.5}$$

And using Eq. (4.3), we obtain:

$$\frac{1}{K'_n \left(\frac{w}{L}\right)_{ne} (V_{DD} - V_T)} = \frac{1}{K'_n \left(\frac{w}{L}\right)_{nA} (V_{DD} - V_T)}$$

$$+ \frac{1}{K'_n \left(\frac{w}{L}\right)_{nB} (V_{DD} - V_T)} \tag{4.6}$$

where $\left(\frac{w}{L}\right)_{ne}$ is the ratio of the width and length for the equivalent or basic NMOS transistor from the basic inverter corresponding with our logic circuit. As the designations A and B for the transistors in the PDN are casual, we can consider these transistors the same. So, these designations, A and B, are indexes and we can write: $i = A, B, C, \ldots$ All the transistors mentioned in Eq. (4.6) are built in the same crystal/wafer; therefore, we may simplify this equation as follows:

$$\frac{1}{w_{ne}} = \frac{1}{w_{nA}} + \frac{1}{w_{nB}} = \frac{1}{w_{ni}} + \frac{1}{w_{ni}} = \frac{2}{w_{ni}} \quad \text{or} \quad w_{ni} = 2w_{ne}. \tag{4.7}$$

The physical meaning of this relation is very simple; we see that two transistors connected in series have two times the width of one equivalent transistor. This is due to the quantity of charged carriers which should transfer through two transistors at the same time as in the one equivalent transistor. If the number of transistors is three, then the width of each transistor will be three times the width of one transistor. Thus, Eq. (4.7) may be generalized as follows for N transistors connected in series:

$$w_{ni} = N w_{ne}. \tag{4.8}$$

Let us consider a parallel connection of the transistors in the PDN as shown in Fig. 4.5. In this case, we must use the equation describing the parallel connections of resistors:

$$R_{PDN} = \frac{R_{nA}R_{nB}}{R_{nA} + R_{nB}} = \frac{\frac{1}{K'_n\left(\frac{w}{L}\right)_{nA}(V_{DD}-V_T)} \cdot \frac{1}{K'_n\left(\frac{w}{L}\right)_{nB}(V_{DD}-V_T)}}{\frac{1}{K'_n\left(\frac{w}{L}\right)_{nA}(V_{DD}-V_T)} + \frac{1}{K'_n\left(\frac{w}{L}\right)_{nB}(V_{DD}-V_T)}}$$

$$= \frac{1}{K'_n\left(\frac{w}{L}\right)_{ne}(V_{DD} - V_T)} = R_{NMOSe}. \tag{4.9}$$

If we assume that the transistors in the PDN are the same, this equation may be simplified as follows:

$$\frac{1}{w_{ne}} = \frac{\frac{1}{w_{nA}} \cdot \frac{1}{w_B}}{\frac{1}{w_{nA}} + \frac{1}{w_{nB}}} = \frac{\frac{1}{w_{ni}} \cdot \frac{1}{w_{ni}}}{\frac{1}{w_{ni}} + \frac{1}{w_{ni}}} = \frac{\frac{1}{w_{ni}}}{2} \quad \text{or} \quad w_{ni} = \frac{w_{ne}}{2}. \tag{4.10}$$

A generalization of the obtained equation for N transistors connected in parallel can be done as follows:

$$w_{ni} = \frac{1}{N}w_{ne}. \tag{4.11}$$

The application of both Eqs. (4.8) and (4.11) enable us to calculate the resistance of any circuit.

In the case of complex circuits, which include both types of connections, connections in parallel and connections in series, we use an additional principle that is named a "worst case" principle. Let us consider the complex logic circuit from Fig. 4.6 once more:

In this circuit, the PDN consists of two branches. Each one includes a different number of transistors. Here, the same state of the circuit is possible under different combinations of input signals, for example the state $F =$ "1" can be reached if:

1. $A =$ "1"
2. $B = C =$ "1"
3. $A = B = C =$ "1"

Evidently, the calculation of the transistor's dimension may be done only once for the circuit. Therefore, we must calculate each branch of the circuit separately (cases 1 and 2). It is clear that the time of operation in the third case will be lower than in case 1 or 2. So, we calculate the dimensions of transistors Q_{NB} and Q_{NC} and after that, the width of the transistor Q_{NA}.

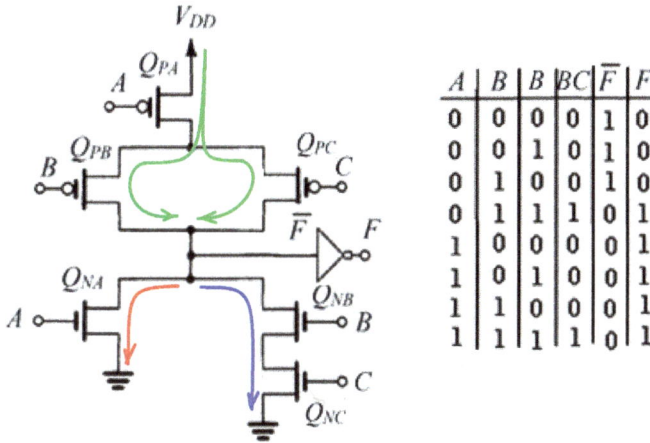

A	B	B	BC	$\overline{F}$	F
0	0	0	0	1	0
0	0	1	0	1	0
0	1	0	0	1	0
0	1	1	1	0	1
1	0	0	0	0	1
1	0	1	0	0	1
1	1	0	0	0	1
1	1	1	1	0	1

Fig. 4.8.   A complex CMOS circuit with signed ways to the current.

To calculate the area required for the circuit, we need values for some initial conditions. These conditions are the length of the transistors, L, the characteristic of its semiconductor properties presented by the relation m_n/m_p, and the relation between the width and length of the basic/equivalent NMOS transistor, $(W/L)_{ne}$. Using the initial conditions, we can calculate the area occupied by the circuit presented in Fig. 4.8.

We can calculate the area occupied by the presented circuit by calculating some of the values associated with the basic CMOS inverter. It consists of two transistors as shown in Fig. 4.9:

The area of this inverter is the sum of the surfaces occupied by the transistors.

$$S_{CMOSe} = S_{PDN} + S_{PUN} = S_{NMOSe} + S_{PMOSe}. \qquad (4.12)$$

The area occupied by the basic NMOS transistor is:

$$S_{NMOSe} = L \cdot W_{ne} = L \cdot \left(\frac{W}{L}\right)_{ne} \cdot L = L^2 \cdot \left(\frac{W}{L}\right)_{ne}. \qquad (4.13)$$

The area occupied by the transistor PMOS in the basic inverter is:

$$S_{PMOSe} = L \cdot W_{pe} = L \cdot \frac{\mu_n}{\mu_p} \cdot W_{ne} = L^2 \cdot \left(\frac{W}{L}\right)_{ne} \left(\frac{\mu_n}{\mu_p}\right). \qquad (4.14)$$

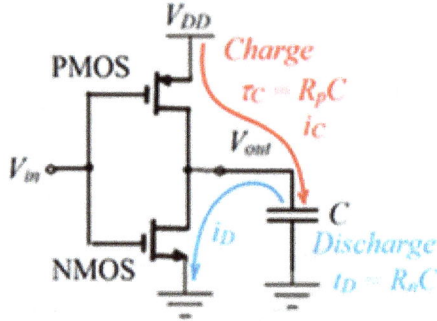

Fig. 4.9. A basic CMOS inverter.

Therefore, the full area occupied by the basic CMOS inverter will be as follows:

$$S_{CMOSe} = L^2 \cdot \left(\frac{W}{L}\right)_{ne} + L^2 \cdot \left(\frac{W}{L}\right)_{ne} \left(\frac{\mu_n}{\mu_p}\right)$$

$$= L^2 \cdot \left(\frac{W}{L}\right)_{ne} \cdot \left(1 + \frac{\mu_n}{\mu_p}\right). \tag{4.15}$$

The calculation of the area occupied by the circuit presented in Fig. 4.8 begins from the PDN, which consists of two branches. The first branch comprises only one transistor Q_{NA}, which provides a path for the current if the input signal A is high. Thus, we can assume that this transistor is equal to the basic NMOS transistor and the area occupied by this transistor is equal to the area occupied by the basic transistor, $S_{na} = S_{NMOSe}$. The second branch of the PDN comprises two transistors connected in series. Therefore, on the basis of our analysis and Eq. (4.8), one can say that the width of both transistors will be two times that of the width of the basic NMOS transistor, $W_{nb} = W_{nc} = 2W_{ne}$. Thus, the area occupied by PDN will be follows:

$$S_{PDN} = S_{na} + S_{nb} + S_{nc} = L^2 \cdot \left(\frac{W}{L}\right)_{ne}$$

$$+ 2 \cdot 2 \cdot L^2 \cdot \left(\frac{W}{L}\right)_{ne} = 5L^2 \cdot \left(\frac{W}{L}\right)_{ne}. \tag{4.16}$$

The PUN part of the circuit consists of three transistors. There are three variations that will produce the low state for the circuit:

1. $A = B =$ "0",
2. $A = C =$ "0",
3. $A = B = C =$ "0".

Evidently, we have here two "worst" cases, 1 and 2, with case 3 having a lower charging time than the "worst" cases. Therefore, we will use the "worst" cases for our calculations. Here, in both "worst" cases we have two transistors connected in series. These transistors may be considered the same as their names were given casually. So, the PUN consists of three equal transistors each of which is two times the basic PMOS transistor. Therefore, the area occupied by the PUN will be:

$$S_{PUN} = S_{pa} + S_{pb} + S_{pc} = 3 \cdot S_{pa} = 3 \cdot 2 \cdot S_{PMOSe}$$

$$= 6 \cdot L^2 \cdot \left(\frac{W}{L}\right)_{ne} \left(\frac{\mu_n}{\mu_p}\right). \tag{4.17}$$

And the area occupied by the circuit presented in Fig. 4.8 will be as follows:

$$S_{CMOS} = 5L^2 \cdot \left(\frac{W}{L}\right)_{ne} + 6L^2 \cdot \left(\frac{W}{L}\right)_{ne} \left(\frac{\mu_n}{\mu_p}\right). \tag{4.18}$$

The described method enables us to calculate the area of some logic circuits built with CMOS technology.

4.2. Dynamic logic circuits

4.2.1. *Principle of operation*

The logic circuits usually work in digital systems initiated with various types of timers. The timers provide the sequence of operations and ensure the implementation of a given logic. Sometimes, a timer divides all the operation time of the digital system into two non-equal and repeating parts: a setup/precharge phase and an evaluation phase. In the first phase, the timer provides a precharge operation for all circuits in a system built with dynamical technology. In the second

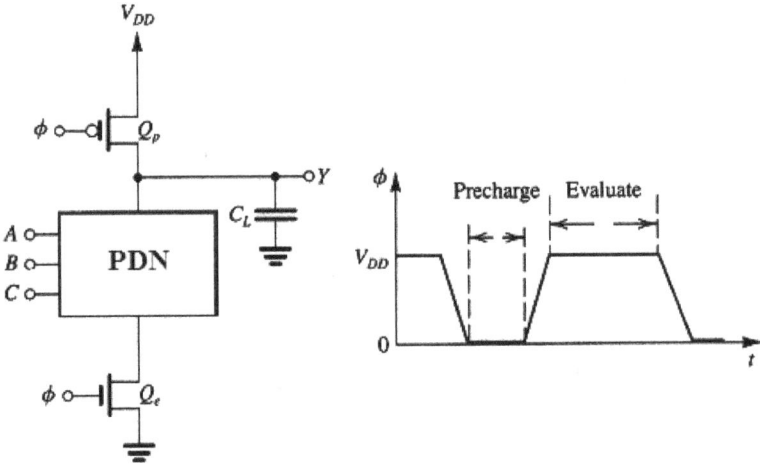

Fig. 4.10.   A dynamic combinational circuit, a general view.

phase, the circuits connected with the systems take a suitable state, "1" or "0". A general view of a digital circuit built with dynamic technology is presented in Fig. 4.10. As shown, this circuit consists of a PDN system only connected with the supply and ground using two specific transistors, Q_p and Q_e. These two transistors are triggered by the timer's signal ϕ and operate in antiphase. This signal divides the circuit's work-time into two non-equal parts: setup/precharge (transistor Q_p) and evaluation (transistor Q_e).

Let us consider the work-cycle of the circuit. The PDN consists of NMOS transistors which have several input variable signals: $A, B, C, \ldots$ Also, the circuit comprises an additional input ϕ which may be at a low or high level. In the case of $\phi = $ "0", the transistor Q_p will be open/conducting and transistor Q_e will be closed. Thus, in the case $\phi = $ "0" the PDN will be disconnected from the ground and the output contact Y of the circuit will be connected with the supply through the transistor Q_p. At this time, the load circuit designated using a capacitor C_L will be charged up to the voltage V_{DD} through the Q_p transistor, $Y = $ "1". The same operation is performed simultaneously in all circuits in the dynamical system. This is the precharge or setup phase of operation.

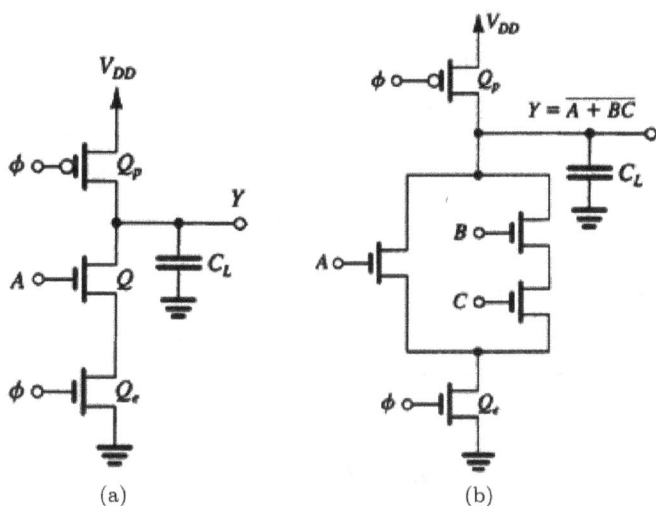

Fig. 4.11.   Examples of simple dynamic logic circuits.

The input signal ϕ changes according to the timing diagram shown in Fig. 4.10. After the growth of this signal, the transistor Q_p will be closed as it is the PMOS transistor and the transistor Q_e will be in the triode state; the evaluation phase of operation begins. Now, the state of the output contact Y will be defined by the logic function implemented by the PDN and the states of the input variables. If all the input variables are in the high state, all the NMOS transistors which are in the PDN will be conductive and the capacitor C_L will be discharged and $Y = $ "0". If the output function is equal to "1", the capacitor will not be discharged and Y will remain the value obtained during the setup phase. In this case, one can say that the rise time of the function is zero.

Fig. 4.11 represents two examples of simple dynamic circuits. A first circuit, shown in Fig. 4.11(a), represents the simplest dynamic circuit, a dynamic inverter. Evidently, the value of the function Y is equal to the inverted value of the variable A. However, we can read this value in the case of $f = $ "1", in the evaluation phase of the operation. The second circuit, Fig. 4.11(b), consists of two transistors Q_p and Q_e ensuring the dynamical type of operation, and the PDN providing a given logic function. A load in both circuits is designated using the load capacitor C_L. Both circuits are dynamical circuits and their work-behavior is divided into two non-equal phases: precharge/setup and evaluation.

We may calculate an area occupied by the dynamical logic circuit using the same principle which was applied to CMOS circuits. In other words, we compare the given dynamic circuit with the basic/equivalent CMOS inverter. The part involving the calculation of the PDN should be done using the "worst" case rule, including in the calculation the evaluation transistor Q_e. The dimensions of the precharge transistor Q_p may be assumed the same as those of the $PMOS_e$ transistor.

4.2.2. *Cascading dynamic logic gates*

Each logic circuit is usually loaded with logic circuits from the same family. Let us consider a serial connection of two of the simplest dynamic circuits as shown in Fig. 4.12. Here one dynamic inverter loads the second dynamic inverter. As both circuits are in the same system, the timer signal f is the same for both inverters, i.e. both circuits have the same setup phase and the same evaluation phase.

The following discourse is a detailed analysis of the system's work:

1. $f = 0$ — setup phase of operation; transistors Q_p are in the triode state and conduct the charge current; the charging paths

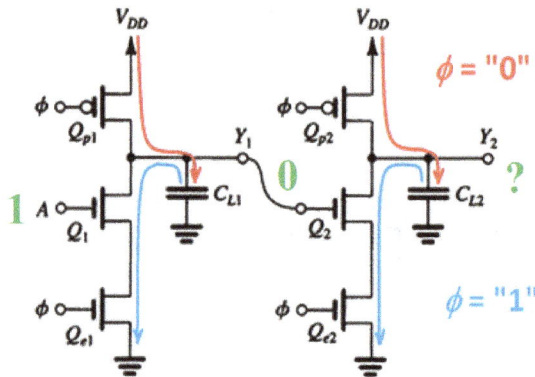

Fig. 4.12. Serial connection of two dynamic inverters.

are colored by red lines; transistors Q_e are in the cutoff state; the value of input variable A is irrelevant, $Y_1 = Y_2 = 1$.

2. $f = 1$ — evaluation phase of operation; transistors Q_e are in the triode state and conduct the discharge current; transistors Q_p are in the cutoff state. In this case, it is possible to have two different behaviors depending on the value of the input variable A:

  a. $A = 0$, transistor Q_1 is in the cutoff state; $Y_1 = 1$; transistor Q_2 is in the conductive/triode state; the capacitor C_{L2} discharges through two transistors Q_2 and Q_{e2}; $Y_2 \Rightarrow 0$; the circuit works properly, from the logic point of view.

  b. $A = 1$, transistor Q_1 is in the conductive state. Due to the initial stage of $Y_1 = Y_2 = 1$, both capacitors C_{L1} and C_{L2} begin discharging simultaneously (the discharge paths are marked by blue lines). When the voltage Y_1 falls lower than the threshold voltage of the second inverter, the output voltage Y_2 also will take time to fall so low that it will be enough to bring the following load from the logic state. Therefore, we obtain the states consequence, indicated by green symbols, which presents the contradiction.

To repair this situation, it will be enough to connect the static CMOS inverter to the output contact of a logic circuit as shown in Fig. 4.13.

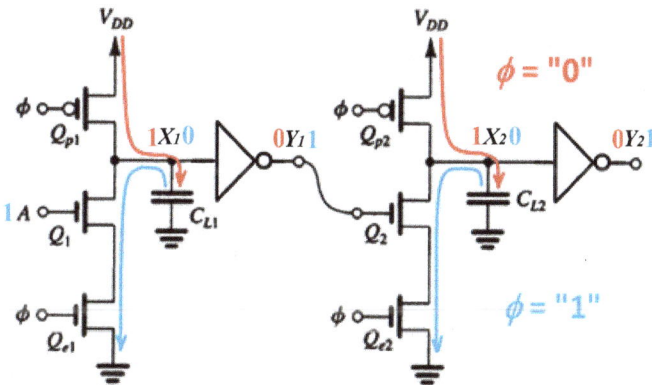

Fig. 4.13.   Improved serial connection of dynamic logic circuits.

Now, the process goes without contradictions. The static inverter introduced between the output contact of the circuit and the input contact of the load play the role of a buffer, which creates a time delay ensuring the reliable operation of the system. In the setup phase, each output contact of the circuit in the system obtains a high logic value from the supply through the precharged transistor Qp. The signal sequence marked by red shows the right logic order. In the evaluation phase, when $f = 1$ and $A = 1$, the capacitors C_{L1} and C_{L2} cannot discharge simultaneously. The transistor Q_2 will begin to conduct a current only after decreasing the voltage in the point X_1 to a value lower than the V_{IH} of the inverter and inverting the Y_1 state from logic 0 (red) to logic 1 (blue). Thus, we will obtain the right order of signals in each contact of the system (marked by blue). The basic CMOS inverter is commonly used as a buffering inverter in the digital dynamic circuits. This type of connection is known in the literature as the "Domino" principle. Here, each loading circuit begins to work after a previous part with the time delay defined by the basic static CMOS inverter.

4.3.   Pass-transistor logic (PTL)

4.3.1.   *The application of PTL to NMOS transistors*

The most convenient logic circuit design uses both the serial and parallel combinations of switches. In this way, one can create various logical functions as shown in Fig. 4.14. Here, the combinations of switches realize such logical functions as logic multiplication

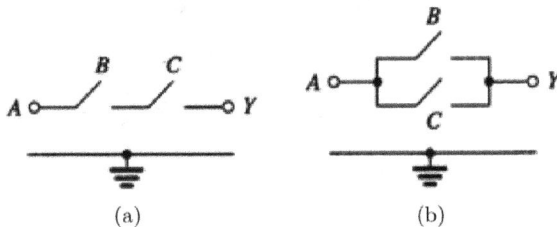

(a)   (b)

Fig. 4.14.   Logical functions designated using switches.

$Y = A \cdot B \cdot C$ (see Fig. 4.14(a)) and the more complex function $Y = A \cdot (B + C)$ (see Fig. 4.14(b)).

A main problem in this type of design is that of finding an appropriate electronic switch which is compact, rapid, and transfers the logic signal without losses. Let us consider an NMOS transistor in the role of the electron switch. It is compact and the signal will go through the open transistor rapidly enough. The propagation time will be minimal in this case. Figure 4.15 represents the NMOS transistor as a switch that enables the transfer of an electrical current in both directions: forward and backward as shown in the figure.

Evidently, current flows to the right side, from A to Y, in the case when $A = $ "1" and $C = $ "1". In the case when $A = $ "0" and $C = $ "1", the current flows to the left side, from Y to A. A transistor which can transfer a current in both directions, forward and backward, is called a "pass transistor". The current's flow process "to the right side" may be illustrated using Fig. 4.16.

This picture explains the pass transistor's behavior. When the input voltage V_A rises to the high state at the time $t = 0$

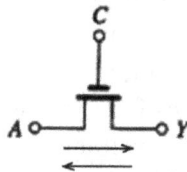

Fig. 4.15.   An NMOS transistor applied as a switch to transmit a carrier current.

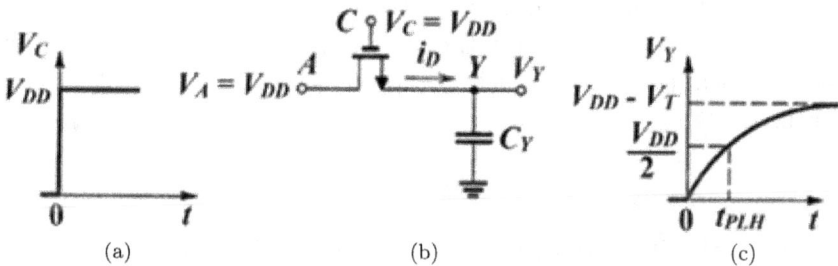

Fig. 4.16.   A right-side current flow case.

(see Fig. 4.16(a)), a current i_D begins to flow through the transistor, as shown in Fig. 4.16(b) and charges the output point Y which is not connected to the ground. The time needed to charge the virtual capacitor C_Y will be equal to the propagation delay t_{PLH} that is calculated upon reaching $V_{DD}/2$ at point Y (see Fig. 4.16(c)).

When $V_{in} = V_A = V_{DD} = $ "1" and $V_C = V_{DD}$, we obtain the NMOS enhanced mode transistor with shortly connected drain and gate. However, the output point Y, which is the source of the transistor, is not connected to ground. A current from the input point A flows through our pass transistor and charges the virtual capacitor C_Y. In this case, the transistor will be in the saturation mode and we can write the following equation for the current:

$$i_D = \frac{1}{2}K_n(V_{GS} - V_T)^2 = \frac{1}{2}K_n(V_C - V_Y - V_T)^2$$

$$= \frac{1}{2}K_n(V_{DD} - V_Y - V_T)^2. \tag{4.19}$$

Therefore, when a current is equal to zero, $V_Y = V_{DD} - V_T$ will be the maximum voltage transferred through our switch. Therefore, we obtain a "poor 1" using an NMOS transistor as the pass-transistor.

In addition, we need to take into account that the output point Y is not connected to the ground, so there is a voltage difference between the body and source of the transistor, $V_{sb} > 0$. This outcome is called the body effect. The voltage that appears due to the body effect influences the threshold voltage and increases it:

$$V_T = V_{T0} + \gamma \left(\sqrt{2\phi_f + V_{sb}} - \sqrt{2\phi_f} \right). \tag{4.20}$$

Where V_{T0} is a threshold voltage in the case $V_{sb} = 0$, Φ_f designates the doping level in the body of the transistor, and γ is the technological parameter which may be calculated using the following formula:

$$\gamma = \frac{\sqrt{2qN_A\varepsilon_0\varepsilon_r}}{C_{ox}}. \tag{4.21}$$

Evidently, if a voltage $V_{sb} = 0$, the threshold voltage will be constant and defined by Eq. (3.13). This body effect may act negatively on the digital circuits. For instance, an output voltage,

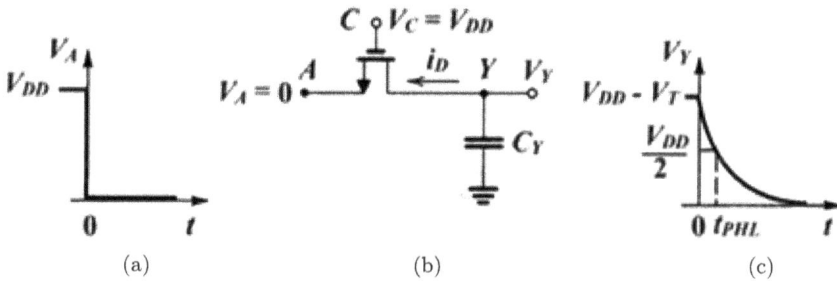

Fig. 4.17. The left-side current flow case.

V_Y, which is an input voltage for the load circuit, may be less than is necessary for the reliable closing of the loading circuit.

Figure 4.17 presents the case of the left flow of the current.

When the voltage at point A drops as shown in Fig. 4.17(a), $V_A = 0$, and a virtual capacitor is full ($V_Y = V_{DD} - V_T$), a discharge current of the capacitor C_Y begins flowing in the left side through the transistor, as shown in Fig. 4.17(b), if the voltage on the gate is equal to "1" ($V_C = V_{DD}$). Figure 4.17(c) illustrates the voltage decreasing process at point Y. Evidently, the processing time here will be equal to t_{PHL} and may be calculated using a suitable equation describing the current flow through the pass transistor. In the case of the left flow, the source and drain of the transistor change places. Therefore, the voltage difference between the transistor's body and the source will be zero, a body effect is absent.

When the input voltage V_A drops, the transistor immediately goes into saturation mode and a current flowing through the transistor will be as follows:

$$i_D = \frac{1}{2}K_n(V_{GS} - V_T)^2 = \frac{1}{2}K_n(V_{DD} - V_T)^2. \qquad (4.22)$$

Now, we obtain a complete discharge of the capacitor C_Y, and at the end of the process, $V_Y = 0$.

Let us consider a series connection of two CMOS inverters through the NMOS pass-transistor. Figure 4.18 represents such a connection. We designate the output of the driving inverter Q_1-Q_2 as A. A switch S_1 represents a logic variable B, which will be equal to "1" in the case of a closed switch and "0" when a switch is open. In other words, the

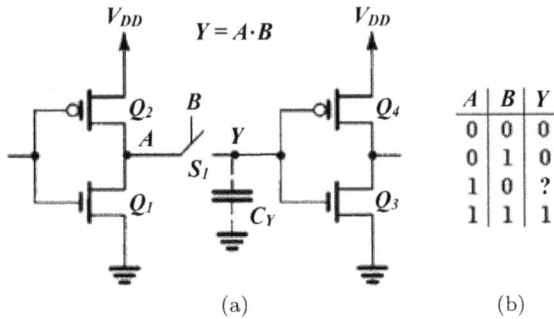

Fig. 4.18.   A series connection of inverters using a pass-transistor.

switch S_1 is a pass-transistor controlled by the signal B. In this case, our circuit implements a logic function $Y = A \cdot B$. Figure 4.18(a) represents an implementation of the function and Fig. 4.18(b) shows the truth table of this circuit.

An analysis of the truth table presented in Fig. 4.18(b) shows that there is an uncertainty in the behavior of the function implemented by our circuit. Here, the virtual capacitor C_Y represents the capacity of the input part of the loading inverter Q_3-Q_4. If the signal B is low (the switch S_1 is open), the high value of the signal A transferred to the point Y cannot be stored and is quickly discharged through the leakage currents. Therefore, the voltage at point Y will be ambiguous and the following circuit will stop working in digital mode, that is, it will come out of logic. This uncertainty illustrates a basic requirement of the connection of digital circuits with pass-transistors: **each electrode and each connection point in the circuit must be connected to the supply or to the ground always.** Thus, the uncertainty in the circuit presented in Fig. 4.18(a) is the result of not supplying or grounding point Y.

To produce the right behavior of the circuit in Fig. 4.18(a), it is necessary to arrange point Y such that it is possible to connect it to the supply or ground. Figure 4.19(a) illustrates one of the schematic solutions to this problem. As can be seen, the switch S_1 in our circuit is equipped with the additional switch S_2 controlled by the complementing signal $\bar{B}$. Using this addition, the point Y will be always defined: if the control signal $B = 1$, the complementing signal

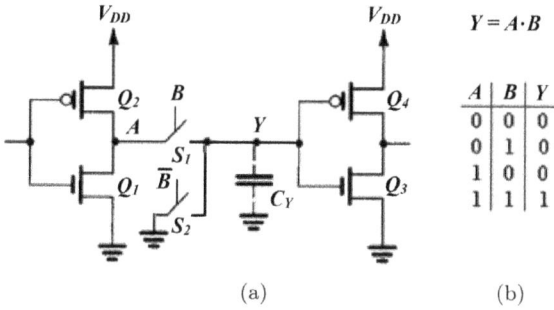

V_{DD}

V_{DD} $Y = A \cdot B$

Q_2 B

A   Y

S_1

Q_1 $\bar{B}$

C_Y

S_2

Q_4

Q_3

A	B	Y
0	0	0
0	1	0
1	0	0
1	1	1

(a)   (b)

Fig. 4.19.   The implementation of multiplication logic with the pass-transistor.

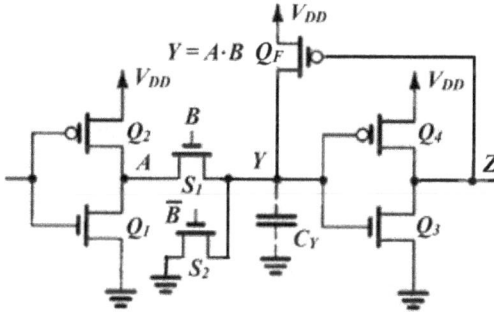

Fig. 4.20.   An additional transistor closes in the scheme of pass-transistor logic.

$\bar{B} = 0$, the switch S_1 is closed and the state of Y is equal to A; if $B = 0$, the complementing signal $\bar{B} = 1$, and $Y = 0$ due to the closed switch S_2. The truth table of the function $Y = A \cdot B$ built for the circuit and working according to the described behavior is shown in Fig. 4.19(b). In this way, the general requirement for the pass-transistor logic circuit will be fulfilled.

Another problem is the decreased voltage transferred through the NMOS pass-transistor. As was mentioned above, the NMOS pass-transistor can transfer only the lowered voltage $V_Y = V_{DD} - V_T$. If V_Y remains constant and V_T increases, it will be possible that the voltage V_Y decreases so much that the following inverter Q_3-Q_4 comes out from the stable logic state. This problem may be solved using an additional transistor of the PMOS type, which will play a supporting role. Figure 4.20 represents this solution.

A circuit with an additional transistor works as follows: when the input signal $A = 1$ and the control signal $B = 1$ also, a virtual capacitor C_Y charges. When the voltage at point Y rises up to the value $V_Y = V_{IL}$ of the inverter Q_3-Q_4, this inverter will turn over in the low state. The voltage $V_Z = 0$ will open the additional transistor Q_F and the point Y will be connected with the supply V_{DD} through the PMOS transistor. Therefore, the virtual capacitor will charge up to the full V_{DD} voltage. As shown, to obtain the maximum voltage transfer through a pass-transistor, we should attach to it another transistor of the PMOS type. NMOS pass-transistors are applied in cases when decreasing of the output voltage is not critical, for example in the dynamical memory circuits that will be considered in Chapter 5.

4.3.2.   *The application of PTL to CMOS transistors*

The problem of decreasing the output voltage in the pass-transistor switches may be solved in another way. Let us connect two complementary transistors, NMOS and PMOS, such that the source of the first transistor will be connected with the drain of the second transistor and vice-versa. Figure 4.21 illustrates such a connection. Here, Figs. 4.21(a) and 4.21(b) represent cases with a current flowing to the right and to the left respectively. Figure 4.21(c) represents the designation of such a connection in the reference literature.

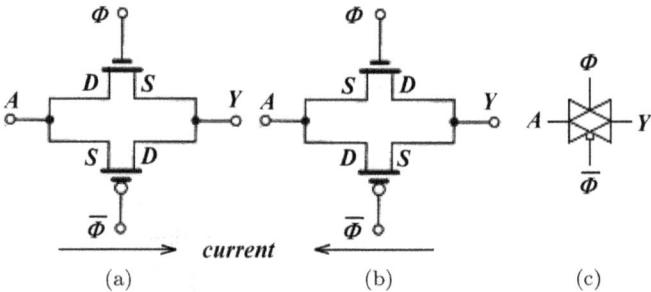

Fig. 4.21.   Two complementary transistors connected as a pass-transistor.

In the circuits shown in Fig. 4.21, we have two different states. When the input voltage $V_A = V_{DD}$ and the control signal $\varnothing = 1$, current flows to the right (see Fig. 4.21(a)). Here, the electrode D, connected with the point A, performs the function of the drain electrode of the NMOS transistor. The source's electrode, S, connects with the output point Y. At the same time, the control signal $\overline{\varnothing} = 0$ ensures the conductivity of the PMOS transistor when the source electrode S is connected to the input point A and the drain electrode D connects with the output point Y. Now, the current flows through both transistors simultaneously. Due to the help of the PMOS transistor, the output voltage reaches up to the maximum possible value, $V_Y = V_{DD}$. If the input point is grounded and the control signals are $\varnothing = 1$ and $\overline{\varnothing} = 0$, the current will flow from the output point Y to point A (see Fig. 4.21(b)). In this case, all the electrodes of the transistors NMOS and PMOS will change their function as shown in the figure. Figure 4.22 represents a schematic view of the technological solution to the parallel connection of two complementary transistors.

This figure looks similar to Fig. 3.39, which described complementary MOS transistors and an elementary invertor built using planar technology. As shown here, the main difference is in the connection of different electrodes. The CMOS transistor can be used to make this connection, placing it in the role of an almost ideal electronic switch, PTL-CMOS.

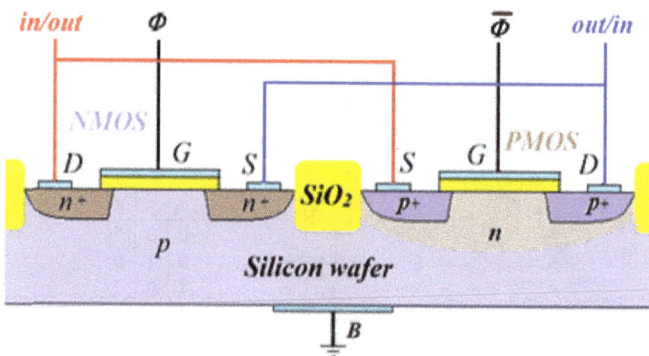

Fig. 4.22.   CMOS pass-transistor, a principal scheme.

4.3.3.   *Combinational circuits using PTL design technology*

As known from Boolean algebra, we only need three basic logic functions to create any logic function of any complexity. These functions are: **AND, OR** and **NOT**. Wherein, these logic functions should be based on our ideal electronic switch, PTL-CMOS. Let us consider a simple scheme to implement the logic function **AND** as shown in Fig. 4.23.

Here, Fig. 4.23(a) represents a structural scheme, Fig. 4.23(b) represents an electrical circuit, and Fig. 4.23(c) shows the truth table of the obtained function. Evidently, if the control signal **B** is high, the output function F will be clearly defined: the voltage at point A and at point F will be equal. However, in the case when $B = 0$, both the NMOS and PMOS transistors in the PTL-CMOS construction will be closed and the value of the voltage at point F became undefined. For example, if before closing the switch, the voltage V_F was high, after a short time this voltage will drop due to leakage currents. Therefore, we must apply something else to implement the logic function.

An interesting variation is to apply a simple multiplexer 2:1 to realize the logic scheme **AND**. Figure 4.24 represents the schematic implementation of the multiplexer 2:1. Figure 4.24(a) shows the structure or functional scheme. The multiplexer transfers the input value V_A or V_B depending on the value of the control signal ϕ to the output point F. When $\phi = 1$, $F = A$ and if $\phi = 0$, $F = B$. This behavior may be described by the following function:

$$F = A\phi + B\bar{\phi}. \qquad (4.23)$$

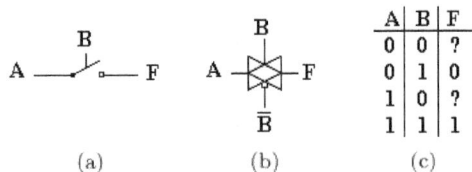

A	B	F
0	0	?
0	1	0
1	0	?
1	1	1

(a)   (b)   (c)

Fig. 4.23.   An attempt to implement the function **AND**.

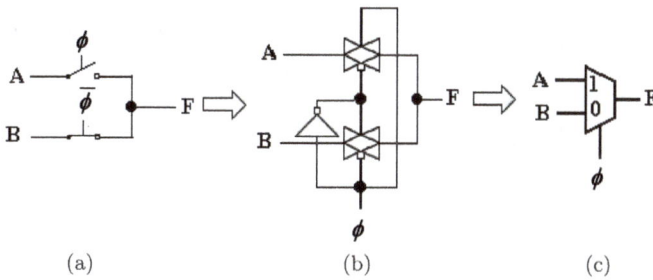

Fig. 4.24.   A schematic implementation of the multiplexer 2:1.

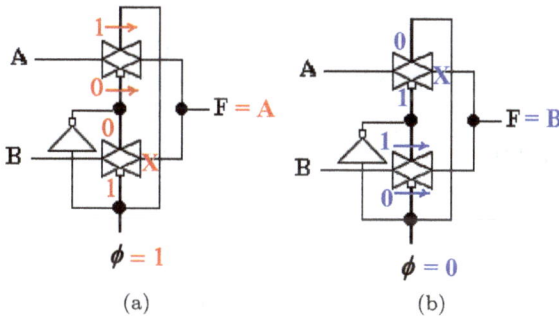

Fig. 4.25.   The principle of operation of the PTL-CMOS implementation of the multiplexer.

Figure 4.24(b) represents a principal electrical circuit implementing the multiplexer 2:1 function. As shown, this circuit contains three different elements: two electronical switches PTL-CMOS (upper and lower) and one elementary invertor. All three elements are built from two complementary MOS transistors. Figure 4.24 presents a schematic designation of the multiplexer 2:1. Figure 4.25 illustrates the operation principle of the circuit.

Assume that the control signal F is high (see Fig. 4.25(a)). In this case, both transistors in the lower switch receive signals closing the transistors on their gates: the PMOS transistor receives the logic "1" and the NMOS transistor receives the logic "0". However, both transistors in the upper switch receive the control signals opening them. Therefore, the input voltage V_A is transferred to the output point F. If the control signal F changes its value to the opposite,

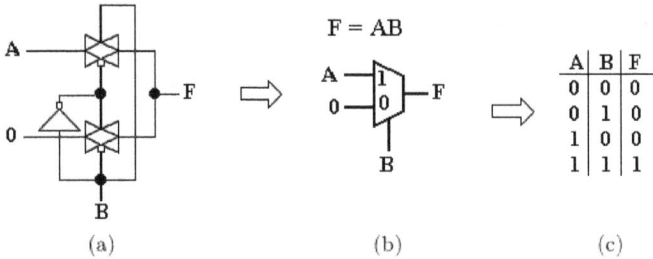

Fig. 4.26.   The PTL-CMOS implementation of the logic function **AND**.

the lower transistors will transfer the input voltage from point B to the output and the upper switch will be open. Using this principle of operation, we can represent the required logic functions according to Eq. (4.23).

Now, let us build the logic function **AND** in the form of Eq. (4.23) using the rules of the Boolean algebra:

$$F = AB = AB + 0 \cdot \bar{B}. \tag{4.24}$$

Figure 4.26 represents the electronic circuit implementing the logic function **AND**. Figure 4.26(a) shows the principal electronic circuit of this function and Fig. 4.26(b) presents a simplified function. As shown here, only a change in the input parameters enables us to implement the logic **AND**.

To prove that the built circuit really implements the required function, Fig. 4.26(c) represents a truth table filled for this circuit. As shown, the obtained circuit truly implements the function **AND**.

The logic and function **OR** may be realized in the same way:

$$F = A + B = A(B + \bar{B}) + B = AB + A\bar{B} + B$$
$$= B(A + 1) + A\bar{B} = 1 \cdot B + A\bar{B}. \tag{4.25}$$

Eq. (4.25) may be implemented in the same way as Eq. (4.24). Figure 4.27 illustrates the electronic circuit on the base multiplexer 2:1 implementing the **OR** function.

As presented in Fig. 4.27, the logic function OR may be realized in two ways: Fig. 4.27(b) presents the logic OR obtained using the complementing through De Morgan's theorem, Fig. 4.27(c)

$F = A + B$ $\overline{F} = \overline{A}\overline{B}$ $F = A\overline{B} + A + B$

A	B	F
0	0	0
0	1	1
1	0	1
1	1	1

(a)   (b)   (c)   (d)

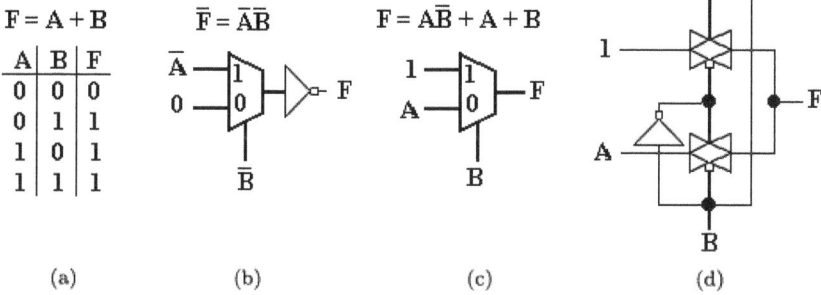

Fig. 4.27. The PTL-CMOS implementation of the logic function **OR**.

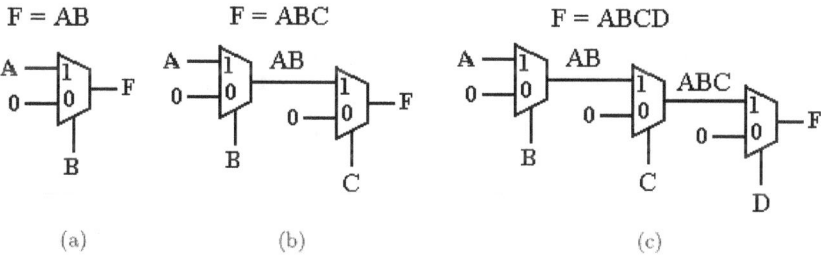

$F = AB$ $F = ABC$ $F = ABCD$

(a)   (b)   (c)

Fig. 4.28. The multiplication of logic circuits on the base PTL-CMOS.

represents the same function in a more logical and optimal way. Sure, the second variation is better as it uses a minimum of various logic elements. Figure 4.27(d) shows the electronic scheme implementing this function. The behavior of this circuit is illustrated by the truth table presented in Fig. 4.27(a).

Now, we have the two basic logic functions **AND** and **OR** realized on the base of the PTL-CMOS and the function **NOT** implemented by the elementary CMOS inverter. Using these functions and their combinations, any complex logic function may be implemented. For example, Figs. 4.28 and 4.29 present the multiplication and addition of logical functions respectively. Evidently, increasing the number of multiplied or added terms significantly influences the propagation time for these circuits. Therefore, they can be efficient in functions with a small number of variables.

Logical functions which are designed using the described technology may use variables of both types: direct and complement. The

$F = A + B$ $F = A + B + C$ $F = A + B + C + D$

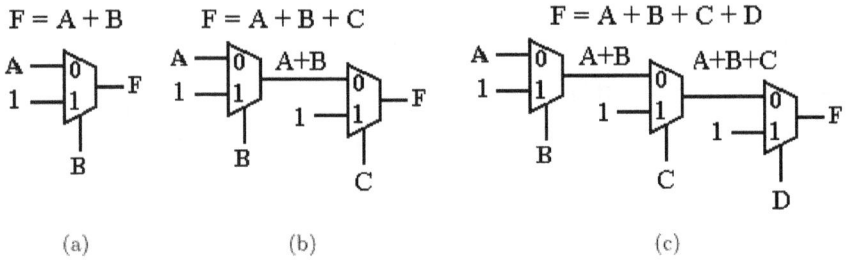

(a)   (b)   (c)

Fig. 4.29.   Logical addition functions on the base PTL-CMOS.

$F = A\bar{B} + \bar{A}B$

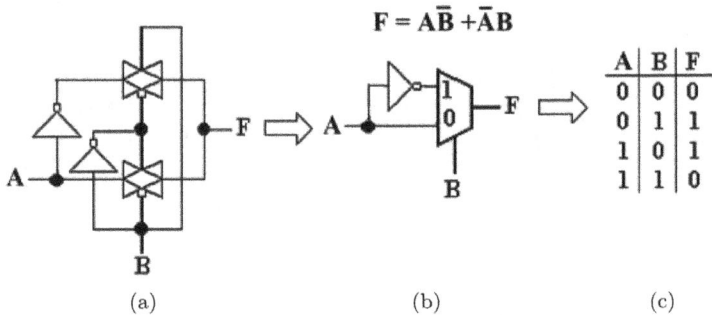

A	B	F
0	0	0
0	1	1
1	0	1
1	1	0

(a)   (b)   (c)

Fig. 4.30.   The PTL-CMOS implementation of the logic function **XOR**.

main problem, bordering on an art, is to bring a given logic function to the form of Eq. (4.23) which allows the use of multiplexers 2:1. Figure 4.30 represents the implementation of logic function **XOR**.

As shown in this figure, the obtained electrical circuit consists of four pairs of CMOS transistors and it is a more effective implementation than the one shown in Fig. 4.7 which was built using CMOS technology. This novel implementation consists of 8 transistors only and it has a minimal area compared with the implementation comprising 12 transistors in Fig. 4.7 and the transistors with doubled areas occupied by each of the transistors in the CMOS scheme. By the way, the calculation of the area occupied by the circuit designed using the PTL-CMOS method is very simple. We need to calculate a number of CMOS pairs in the circuit and multiply this number by the area of the elementary (basic) CMOS inverter.

4.4. Sequential circuits

4.4.1. *Latch and RS flip-flops*

All the circuits considered in previous parts were combinational circuits. Now, we begin to study the sequential circuits. Figure 4.31 represents an example of such a circuit. The state of the outputs in the sequential logic circuit depends on the status of the input signals and the previous state of the output signals. Therefore, each sequential circuit comprises one or more feedbacks and depends on the previous state of the circuit. In other words, the sequential circuit always consists of feedback or memory elements and a combinational environment.

To understand how the sequential circuit is built and how it works and what is a memory element, let us consider the simplest circuit consisting of two inverters when the output of the circuit is connected with the input as shown in Fig. 4.32. The main property of this circuit is the stable state which is retained for a very long time. After connection to the supply, this circuit obtains any state, "0" or "1" and keeps it the same. Such a circuit is called a **latch**. Therefore, a latch is the minimal memory element. This memory element may be in two opposite states and is called a bi-stable element or a static sequential circuit.

The latch circuit shown in Fig. 4.32 uses general designations of inverters, therefore we can use any inverter technology instead of this

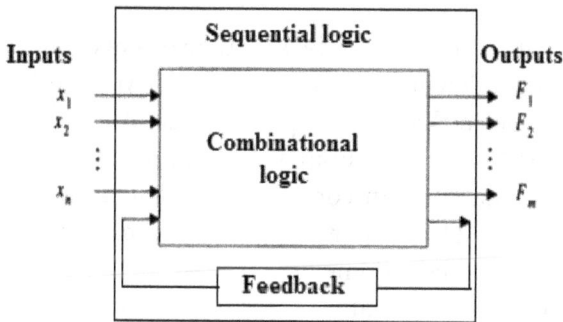

Fig. 4.31.   A general definition of the sequential logic circuit.

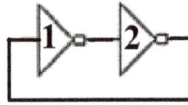

Fig. 4.32.   The latch circuit.

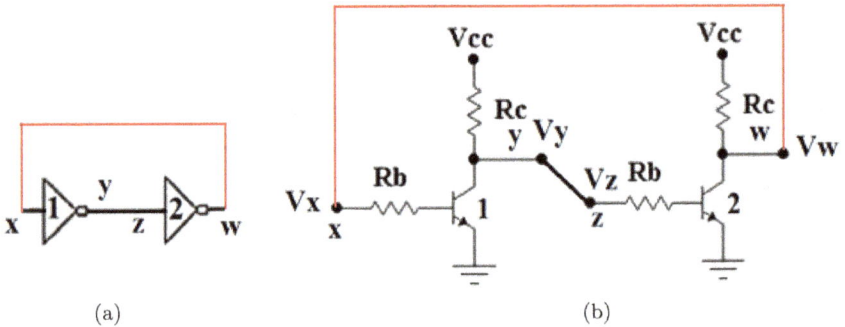

(a)   (b)

Fig. 4.33.   An open circuit scheme of the latch.

general designation. If we assume that the transfer characteristics of both inverters are not changed, let us open the circuit shown in Fig. 4.32. Figure 4.33(a) represents a two-inverter latch circuit, with a red line showing the connection which we open. Figure 4.33(b) represents the same circuit; however, the inverters here are built using RTL technology. Moreover, all the elements marked by identical signs have the same parameters and all the transistors used are identical. The red line has the same function as in the general scheme, Fig. 4.33(a).

The open circuit consists of two RTL inverters connected in series, behaving in the following way: when an input voltage V_x is high, the first inverter inverts this value and V_y is low. As $V_y = V_z =$ "0", the second inverter inverts this input voltage and the output voltage V_w will be high so that we can connect this point, w, with the point x (see the red line). This circuit, Fig. 4.33(b), may be rearranged as shown in Fig. 4.34. The red line here represents the positive feedback.

This scheme has two stable states. It works as follows: after the circuit is supplied by a supply voltage V_{CC}, it exists in the unstable equilibrium state. Now, the base current of transistor 1 is equal

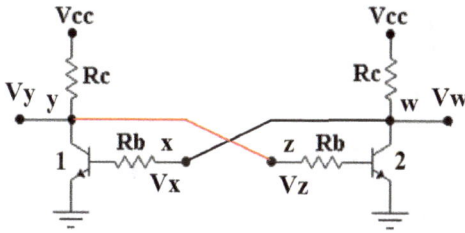

Fig. 4.34.   The rearranged principal electrical scheme of the latch.

to the base current of transistor 2 and both collector currents are equal also. However, the circuit cannot be in this state for a long time. Assume that in a casual way, the base current of the second inverter has increased a little bit. Due to this change, transistor 2 will open a little more and the collector current of this transistor will increase also. The voltage drop on the resistor R_c of the second inverter will grow and the potential of the point w, V_w, will decrease. As $V_w = V_x$, transistor 2 will close up slightly and the collector current of this transistor will decrease. So, the voltage drop on the R_c resistor of the first inverter will be reduced and the potential V_y at the point y will increase more. The process described resembles an avalanche. After the end of this avalanche process, the circuit will be in the stable state when transistor 2 is in the saturation mode and transistor 1 is in the cutoff state. This state is stable and the circuit cannot leave it. To change the state of the circuit, we need to attach additional electrodes as shown in Fig. 4.35. This circuit is called an SR-latch.

Now, we may take suitable designations for the electrodes. We designate the two output points w and y as the states of the circuit Q and $\bar{Q}$ respectively. These values are always opposite to one another. The input connectors, S and R, are designated by their functions **Set** and **Reset**, respectively. The function **Set** ($S =$ "1") is an action that brings the circuit to the stable state $Q =$ "1". The function **Reset** ($R =$ "1") acts in the opposite way. The analysis of the circuit behavior described above brings our logical circuit to the state $Q = 0$ and $\bar{Q} = 1$. To change this state, we need to take $S = 1$ and $R = 0$. Moreover, when conducting an analysis of such circuits, we need to

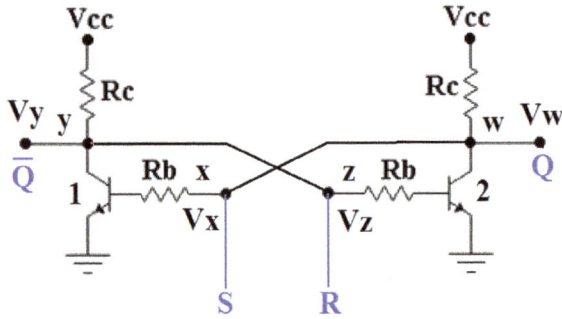

Fig. 4.35.   The simplest SR-latch circuit.

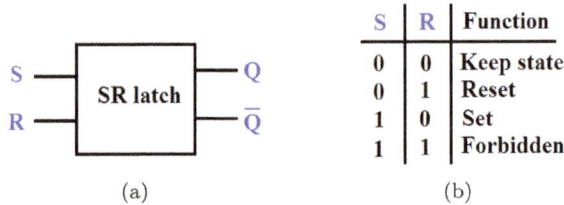

S	R	Function
0	0	Keep state
0	1	Reset
1	0	Set
1	1	Forbidden

(a)   (b)

Fig. 4.36.   The general logical scheme and functional table of the SR-latch.

always define the initial state of the circuit. If $S = 1$ and $R = 0$, the transistor 1 switches to the saturation mode of operation, the output Q will be equal to the logical 1 and $\bar{Q}$ will be low together with the input $R = 0$. A general logical scheme of the latch is presented in Fig. 4.36(a).

As shown in Fig. 4.36(a), the latch-circuit may be represented by a quadrangle with two input electrodes and two complimentary outputs. This device works according to the functional table presented in Fig. 4.36(b). On the base of this functional table, we may build the truth table of the latch as shown in Fig. 4.37(a).

The simplest way to obtain a logic function in its analytical form on the base of a truth table is presenting the truth table in the form of a Karnaugh map (see Fig. 4.37(b)). The combination of neighbor function values enables us to simplify a final function:

$$Q_1 = Q_0 \bar{R} + S \bar{R}. \qquad (4.26)$$

Q	S	R	$\bar{Q}$
0	0	0	0
0	0	1	0
0	1	0	1
0	1	1	X
1	0	0	1
1	0	1	0
1	1	0	1
1	1	1	X

(a)

Q \ SR	00	01	11	10
0	0	0	X	1
1	1	0	X	1

(b)

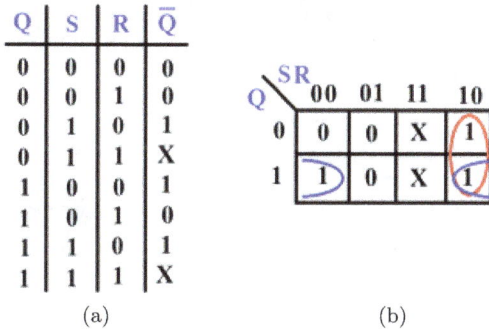

Fig. 4.37. True-table and a Karnaugh map of the latch.

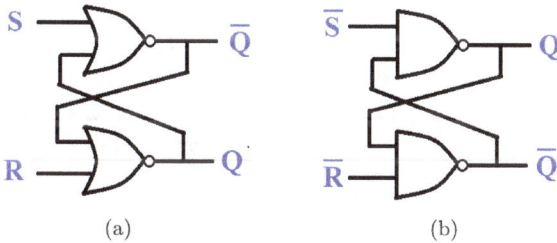

(a)   (b)

Fig. 4.38. A logic schemes of the latch with implementation using NOR and NAND logic gates.

A simple transformation of this logic function presents it in the form of two connected NOR logic functions:

$$Q_1 = Q_0\bar{R} + S\bar{R} = (Q_0 + S)\bar{R} = \overline{\overline{(Q_0 + S)} + R}. \qquad (4.27)$$

Therefore, the latch function will look as presented in Fig. 4.38(a). Here, two NOR logic gates are connected according to Eq. (4.27). A feedback signal Q_0 participates in the appearance of the following signal Q_1. Now it is clear that the electrical circuit shown in Fig. 4.34 is a special case of the general circuit shown in Figs. 4.36(a) and 4.38(a). Figure 4.38(b) represents the same function; however, it is built using NAND gates.

Let us consider the circuit shown in Fig. 4.38(a) once more. Now, we put into the logic gates NOR their implementation built using CMOS technology. We build two NOR circuits and connect them in a ring shape as presented in Fig. 4.39.

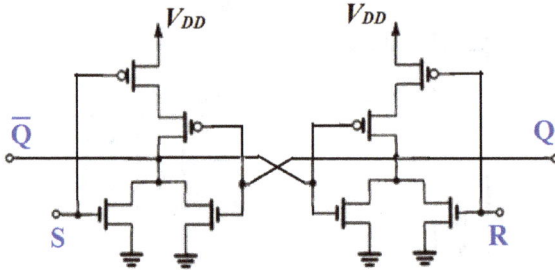

Fig. 4.39.   A direct CMOS implementation of the latch S-R circuit.

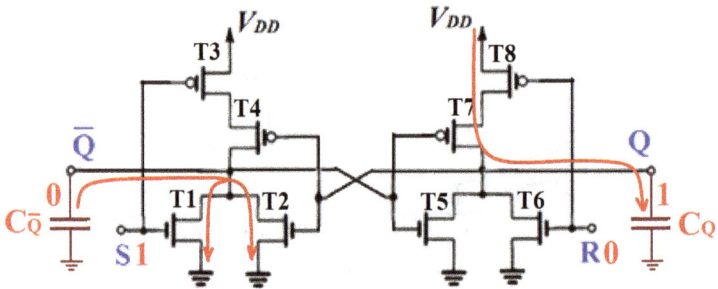

Fig. 4.40.   The latch circuit is in the high logic state.

As was mentioned above, to analyze the sequential circuit we need to take into account the initial stages of the inputs and outputs. Assume that our circuit is in the high logic state and that it has the following input signals: $S = 1$ and $R = 0$ as shown in Fig. 4.40. Signal $S = 1$ (Set) opens the transistor T1 and closes the transistor T3. This leads to the discharging of all charges which were on the virtual capacitor $C_{\bar{Q}}$ that is colored red. The high state of the circuit ($Q = 1$) opens the transistor T2 and closes the transistor T4.

Thus, at the left NOR gate, transistors T1 and T2 are open and transistors T3 and T4 are closed. Therefore, the capacitor $C_{\bar{Q}}$ discharges through both transistors T1 and T2. On the other hand, initial state $R = 0$ together with $\bar{Q} = 0$, closes transistors T5 and T6 and opens transistors T7 and T8. So, the virtual capacitor C_Q will be shortly connected with the supply V_{DD} and the charges through transistors T7 and T8, as shown in Fig. 4.40. After charging

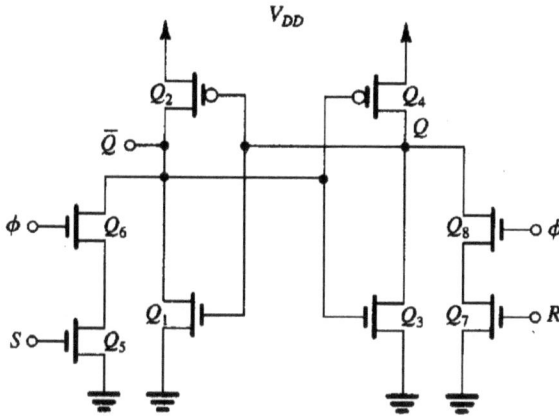

Fig. 4.41.   A synchronous latch or flip-flop based on two inverters.

and discharging suitable capacitors, the latch circuit enters into the stable state and remains in it until it receives the novel control signals S and R. The novel control signals will change the state of the latch. This circuit is good and behaves according to its assignment. However, this circuit contains two deficiencies: the first one is that there are eight transistors in the circuit, and the second is that this circuit is asynchronous, it does not work according to the timer.

The circuit presented in Fig. 4.41 represents a synchronous latch or flip-flop triggered using a timer which generates control signals f.

This circuit represents an integration of two different technologies: CMOS-latch technology and pass-transistor logic. Here, there are two pairs of PTL-NMOS transistors: Q_5-Q_6 and Q_7-Q_8. If the signal f is zero, the circuit retains their state and all changes are forbidden due to the cutoff state of both transistors Q_6 and Q_8. If the signal f is high, the transistors Q_6 and Q_8 will be in the triode state that enables them to change the state of the circuit using external signals S and R. This circuit does not consume power in the stable states.

Let us consider the work of the synchronous latch shown in Fig. 4.41. Assume that the initial state of the circuit is Reset: $Q = 0$, $\bar{Q} = 1$, $V_Q = 0$, $V_{\bar{Q}} = V_{DD}$. Now, we want to set this circuit to the

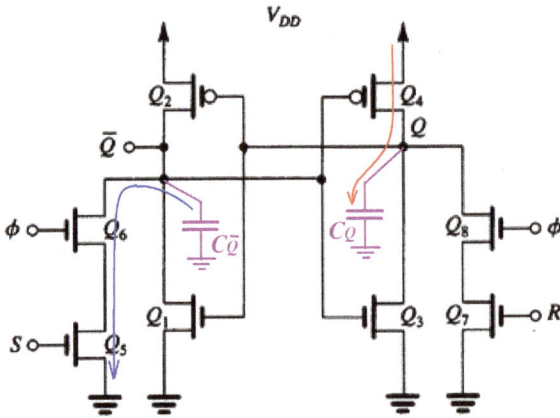

Fig. 4.42.   An S-R flip-flop based on the behavior of the two inverters.

state Set. To do this, we put S = "1" and R = "0". Figure 4.42 illustrates the switching process.

Here, the virtual capacitors represent the input capacity of both inverter circuits Q_1-Q_2 and Q_3-Q_4. In the initial state, the capacitor $C_{\bar{Q}}$ is charged for V_{DD} and second capacitor C_Q is fully discharged. Evidently, the change is possible only for signal f = "1". In our case, both the transistors Q_5 and Q_6 will conduct a discharge current and the voltage at the point $\bar{Q}$ will drop to zero. When this voltage reaches the threshold value of the inverter Q_3-Q_4, it will switch and the virtual capacitor C_Q will be charged up to V_{DD} through the transistor Q_4. In this way, the circuit will shift to the state Set (Q = "1"). This high state $Q = V_{DD}$ will open the transistor Q_1 and close the transistor Q_2. Thus, the point $\bar{Q}$ will be reliably connected to ground and will provide $\bar{Q}$ = "0"; the switching process will be finished.

The same behavior may be realized by excluding the transistors Q_5 and Q_7 from the circuit. A simplified version of the S-R flip-flop is presented in Fig. 4.43.

A very popular circuit realizing an S-R flip-flop is shown here. Static computer memories of small dimensions are usually based on this circuit. The two inverters' latch is synchronized with a timer by the signal f and the two pass-transistors Q_5 and Q_6 are applied here.

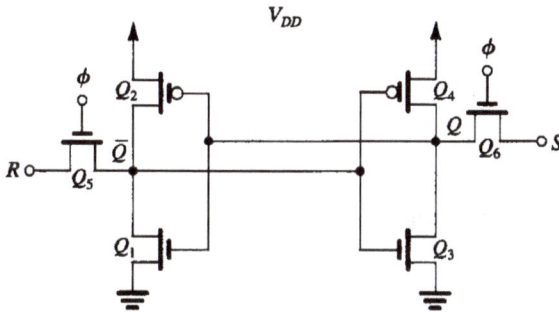

Fig. 4.43. A simplified *S-R* flip-flop circuit.

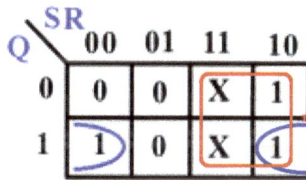

Fig. 4.44. The rearranged Karnaugh map for the *S-R* flip-flop.

The *S-R* flip-flop may be implemented using PTL-CMOS technology also. Let us consider once more the Karnaugh map from Fig. 4.37(b) rearranged as shown in Fig. 4.44.

Here, we rearranged the neighboring variables and obtained the following functional equation of the *S-R* flip-flop: $Q_{t+1} = S + Q\bar{R}$. Using the implementation of AND and OR logic functions by PTL-CMOS technology, the *S-R* flip-flop will take the shape presented in Fig. 4.45.

4.4.2. *Other types of flip-flops*

Let us consider Fig. 4.43 once more. To avoid the insertion of two identical signals $S = R =$ "1" in the circuit and to increase the reliability of the circuit, we can connect the inputs through an additional inverter as shown in Fig. 4.46.

Figure 4.46 represents a so-called *D* (data) flip-flop. This device is intended for storing and transmitting information. Figure 4.46(a)

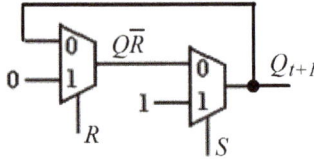

Fig. 4.45.   The *S-R* flip-flop implementation using PTL-CMOS technology.

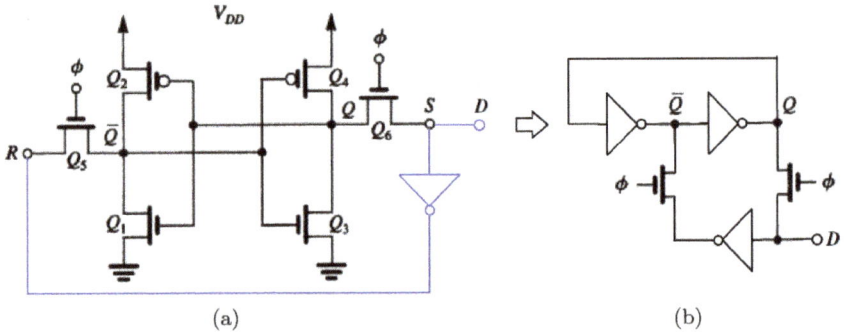

Fig. 4.46.   A *D* flip-flop circuit.

represents an *S-R* flip-flop with the inputs inversely connected through an inverter. Figure 4.46(b) shows a general view of the synchronized *D* flip-flop. When the external signal $f = 1$, it enables us to insert information into the *D* flip-flop. This time, the flip-flop obtains the data and $Q = D$. When the synchronic signal f is zero, the flip-flop stores the information.

The same *D* flip-flop may be realized in the simplest way using only two inverters as shown in Fig. 4.47.

In this circuit, a synchronic signal f and a non-overlapping signal $\bar{\phi}$ are controlling the flip-flop behavior. When $f =$ "1" and $\bar{\phi} =$ "0", novel information, *D*, can enter into the flip-flop. When $f =$ "0" and $\bar{\phi} =$ "1", the stored data is kept inside the flip-flop. The flip-flops may be organized into registers. A line of *D* flip-flops may be used for parallel storage and the transmitting of information in various digital systems such as computers, for example. The *D* flip-flops may be more complicated than shown in Fig. 4.46. Thus, the *D* flip-flop called "master-slave", shown in Fig. 4.48.

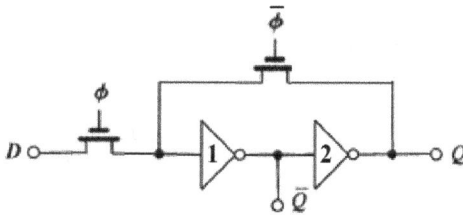

Fig. 4.47.   The simplest D flip-flop based on two inverters.

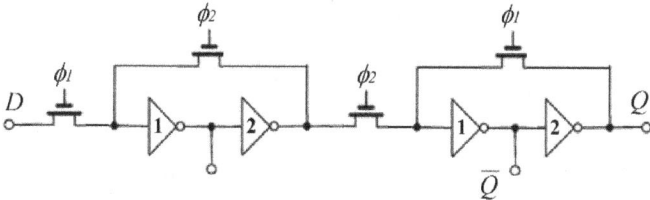

Fig. 4.48.   The "master-slave" D flip-flop.

This type of D flip-flop consists of two identical cells, one of which represents the simplest D flip-flop. The main requirement of this circuit is the non-overlapping of control signals f_1 and f_2. This circuit works in two stages. When $f_1 =$ "1" and $f_2 =$ "0", novel information D enters the first cell of the flip-flop. The second cell now keeps the old information and returns it from the output of the first inverter of the cell. If the control signals change their values, the novel information enters the second cell and the first cell will store this D also.

A D flip-flop may be implemented using PTL-CMOS technology also. As the functional equation of the D flip-flop is $Q_{t+1} = D$, the D flip-flop will be realized as shown in Fig. 4.49.

As shown, this circuit was implemented in the simplest way. The first implementation, presented in Fig. 4.49(a), represents a system storing a piece of inserted information. Usually, several D flip-flops arranged together represent a so-called register used for the storage and transmission of information. The second implementation, shown in Fig. 4.49(b), represents a synchronized D flip-flop. Here, the output signal stays the same at the time $T = 0$ and enables changes in the D flip-flop when $T = 1$.

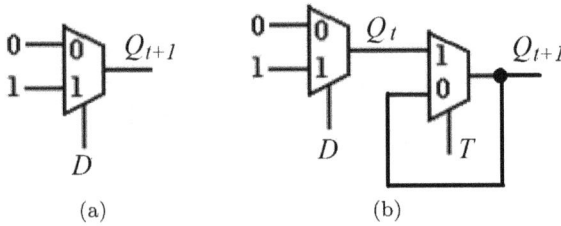

Fig. 4.49.   The D flip-flop implementation. (a) The D flip-flop and (b) the synchronized D flip-flop.

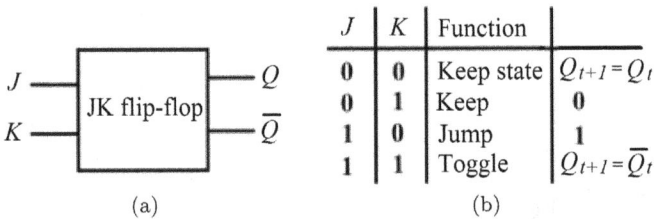

J	K	Function	
0	0	Keep state	$Q_{t+1} = Q_t$
0	1	Keep	0
1	0	Jump	1
1	1	Toggle	$Q_{t+1} = \overline{Q}_t$

(a)   (b)

Fig. 4.50.   The general logical scheme and functional table of the JK flip-flop.

Another type of flip-flop is the well-known J-K flip-flop. A general logical scheme and functional table of the JK flip-flop is presented in Fig. 4.50.

As shown in Fig. 4.50(a), the JK flip-flop may be represented by a quadrangle with two input electrodes and two complimentary outputs. This device works according to the functional table presented in Fig. 4.50(b). One can see that this flip-flop has an additional function compared with the SR flip-flop. Thus, this device can work as a generator of a square wave in the state called "Toggle". On the base of this functional table, we may build the truth table of the JK flip-flop as shown in Fig. 4.37(a). Figure 4.51 represents the truth table and the Karnaugh map for the JK flip-flop.

The simplest way to obtain a logic function in its analytical form on the base of a truth table is the presentation of the truth table in the form of a Karnaugh map. The combination of neighbor function values enables us to simplify a final function:

$$Q_{t+1} = Q_t \bar{K} + J \bar{Q}_t. \qquad (4.28)$$

Q_t	J	K	Q_{t+1}
0	0	0	0
0	0	1	0
0	1	0	1
0	1	1	1
1	0	0	1
1	0	1	0
1	1	0	1
1	1	1	0

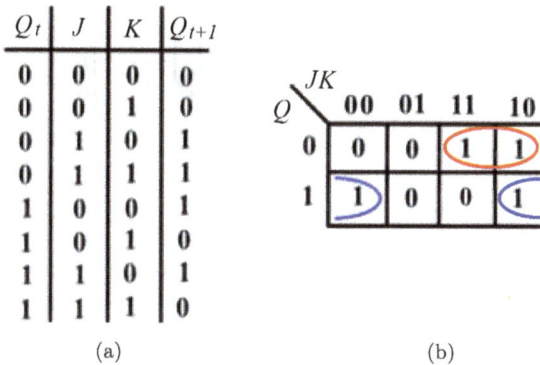

(a)   (b)

Fig. 4.51.   The truth table and the Karnaugh map for the JK flip-flop.

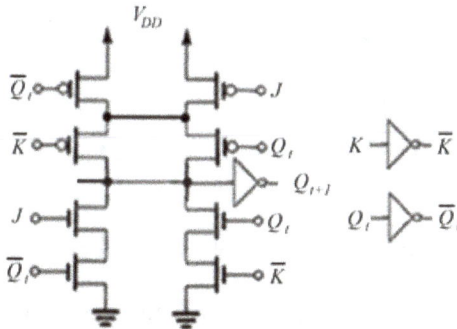

Fig. 4.52.   The CMOS implementation of a JK flip-flop.

This function can be directly implemented using CMOS technology. To achieve this goal, we may build the PDN part of the circuit using Eq. (4.28). The PUN part of the circuit will be obtained using De Morgan's theorem. Figure 4.52 represents the CMOS implementation of a JK flip-flop.

As shown, this system represents a device with enclosed feedback. The current output signal value defines each following value of the output signal unambiguously. The JK flip-flop may be implemented using PTL-CMOS technology also. A simple implementation is shown in Fig. 4.53.

If we put $J = K = 1$, as shown in the functional table of the JK flip-flop (see Fig. 4.50(b)), we will obtain another type of flip-flop,

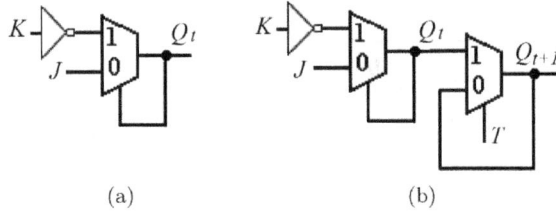

(a)   (b)

Fig. 4.53. The PTL-CMOS implementation of the JK flip-flop.

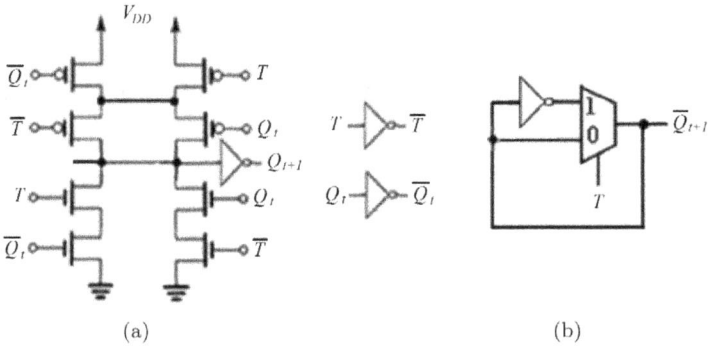

(a)   (b)

Fig. 4.54. The PTL-CMOS implementation of the T flip-flop.

the T flip-flop. Let us consider a function presented in Eq. (4.28) in these conditions:

$$Q_{t+1} = Q_t\bar{K} + J\overline{Q_t} = Q_t\bar{T} + T\overline{Q_t} = Q_t\bar{1} + 1\overline{Q_t} = \overline{Q_t}. \qquad (4.29)$$

Therefore, this function shows that the output signal in the T flip-flop changes its value on the opposite, for example logic "1" changes on the logic "0". The circuit implementation of this flip-flop with CMOS technology and PTL technology will be simplified. The principal electrical circuits of both implementations are presented in Fig. 4.54.

Here, Fig. 4.54(a) presents the CMOS implementation of a T flip-flop and Fig. 4.54(b) shows the PTL-CMOS implementation.

4.4.3. *Dynamic shift register*

We considered four types of basic flip-flop circuits applicable as feedback elements in the sequential circuits. In general, a sequential

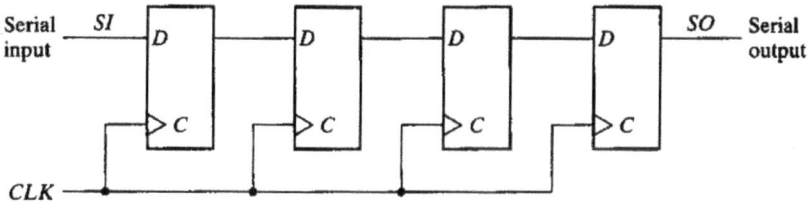

Fig. 4.55. The shift register.

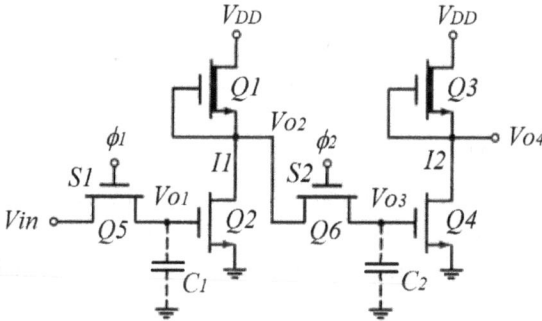

Fig. 4.56. The two-stage dynamical shift register.

circuit represents a set of various flip-flops connected together so that every flip-flop can keep and transfer a bit of information. Such a circuit is called a **register**. Usually, a register consists of the same type of flip-flops and a combinational part. If a register can change the binary states of the flip-flops according to a defined order, it is called a **counter**.

Figure 4.55 represents a schematic view of a specific type of register: the shift register. It has many applications in digital circuit systems.

This shift register may be realized using various technologies. For example, Fig. 4.56 shows a two-stage dynamical shift register implemented by the integration of NMOS-PTL technology together with NMOS technology.

The circuit presented in Fig. 4.56 consists of two identical cells constructed from an NMOS-PTL switch and an NMOS inverter with a depletion-type load. Also, the virtual capacitors C_{L1} and C_{L2} are shown in the circuit. These capacitors represent the input capacity of suitable inverters. Two external timing signals F_1 and F_2 control this circuit. A main requirement of these timing signals is that they do not overlap. To understand the operational principle of this shift register, let us consider the timing diagram presented in Fig. 4.57. In the initial state, t_0, the input signal V_{in} is zero, both the control signals F_1 and F_2 are also zero, both the switches $S1$ and $S2$ are open, the signal V_{01} is zero and the suitable signal V_{02} is in the high state, $V_{02} = 5$ V. Both signals V_{03} and V_{04} are not defined now. The input signal changes its value from "0" to "1" at the time t_1. The first change in the circuit occurs at the time t_2 when the first control signal F_1 opens the transistor $Q5$. This time, the capacitor C_1 will be charged up to "1" ($V_{DD} - V_T = 4$ V) and in accordance with this, the voltage V_{02} decreases to zero after inversion by the inverter $I1$. At the time t_3, the input signal falls to zero, however the circuit cannot respond to this due to the closed switch $S2$. The switch $S2$ opens at the time t_4. Now, the signals V_{01} and V_{02} are still the same, however the signals V_{03} and V_{04} begin to be defined, as $V_{03} = V_{02} = 0$ and the inverter $I2$ inverts this voltage to the high level, V_{DD}. At the time t_5, we can see that the input signal is zero and the voltages V_{01} and V_{02} are returned to their initial states. After opening the switch $S2$ at the time t_6, the input signal comes to the second cell and we can measure the states $V_{03} = 4$ V ("1") and $V_{04} = 0$ ("0"). So, we can see how the input signal transfers through the shift register.

The same logic function may be implemented using advanced technologies as shown in Fig. 4.58. Here, the shift register is also built from two identical cells, however they are implemented using an integration of CMOS and PTL-CMOS approaches. This shift register operates in the same way as the one shown in Fig. 4.57.

Evidently, the discharge time of the virtual capacitors C_1 and C_2 must be more than the operation period of the control signals F_1 and F_2.

4.4.4.   *Multivibrator circuits*

All the sequential circuits considered above include feedback elements implemented as flip-flops of various types. These flip-flop circuits are the two-stable states or bistable systems, as the output signal in these

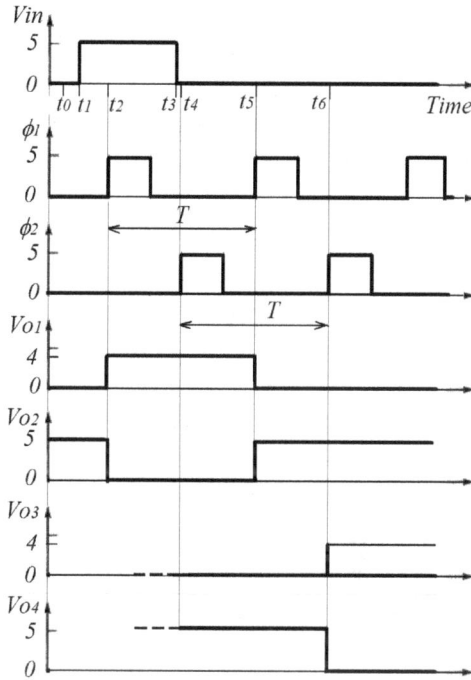

Fig. 4.57. Timing diagram of the shift register.

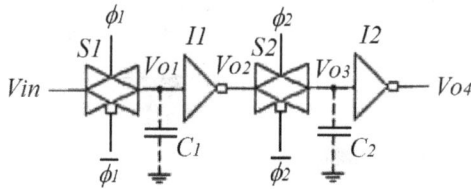

Fig. 4.58. The two-stage dynamical shift register based on CMOS technology.

circuits can be in two alternative states: high or low level (logic "1" or logic "0"). A bistable circuit represents a family of circuits called **multivibrators**. Multivibrators may be one of three types: bistable, monostable, and astable.

4.4.4.1. *Monostable multivibrator*

A monostable multivibrator circuit has one stable state only. This circuit may be in the stable state for a long time. Also, this circuit may be switched to the second, quasi-stable state, in which the circuit will be defined for a limited time. Therefore, one can say that this circuit generates an impulse of defined length, T. Thus, this circuit may be called a "one-shot" or a pulse-stretcher or a pulse-standardizer. The principle of operation of a monostable multivibrator is illustrated in Fig. 4.59.

A pulse duration, T, in the monostable multivibrator, is defined by the internal arrangement of the circuit. So, it is independent from the input pulse duration. Each change of the circuit's state requires an energy storage element. We know two types of such elements: the inductivity coil and the capacitor. The main limitation of the inductivity coil is its volume; this element is a three-dimensional element. So, it cannot be used in integrated circuits built with planar technology. Therefore, we will only study the circuits based on capacitors. By the way, elements of a passive circuit such as resistors and capacitors may be easily implemented using unipolar transistors.

Figure 4.60 represents a monostable multivibrator circuit implemented using two NOR logic gates.

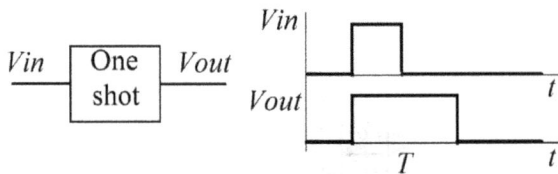

Fig. 4.59.   The principle of operation of a one-shot circuit.

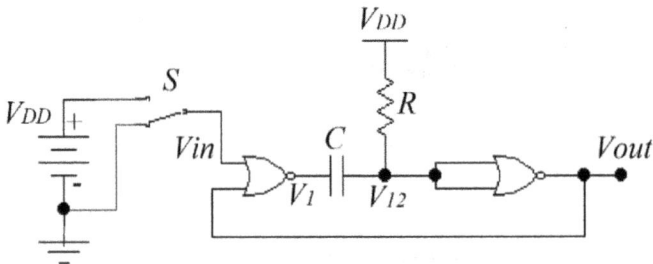

Fig. 4.60.   The principal scheme of the monostable multivibrator.

This circuit consists of two logic NOR elements, a capacitor and a resistor. The first problem with this type of circuit is that of defining the stable state of the circuit. Evidently, in the stable state, all transport processes should be stopped. In other words, all currents in the circuit must be zero at the stable state. The point V_{12} in the circuit is connected to the V_{DD} supply through a resistor R. Therefore, in the steady state, the voltage V_{12} is equal to V_{DD} or "1". To keep the current passing through the capacitor at the level zero, the potential V_1 should be in the high logic state "1" also. Thus, we must keep the input voltage V_{in} and the output voltage V_{out} at the zero level in the stable state. So, the stable state or initial state in the circuit shown in Fig. 4.59 involves the following parameters:

$$V_{in} = V_{out} = 0 \quad \text{and} \quad V_1 = V_{12} = 1.$$

When the switch is turned on in the state $V_{in} = V_{DD}$, the first NOR gate inverts the state and V_1 will return abruptly to the low state. This change looks like the δ-function. As the resistance of the capacitor is inversely proportional to the frequency of the voltage change, $R_c = 1/\omega C$, in the case of the δ-function this resistance tends to zero. Sometimes, it is said that the capacitor is transparent to the high-frequency pulses. Thus, voltage $V_{12} = V_1 = 0$ and the second NOR circuit, which is connected as the NOT circuit, also inverts. In this way, the output voltage will jump to the high level, $V_{out} = V_{DD} =$ "1". At the same time, the charging process of the capacitor will begin through the following elements of the circuit: **Supply $V_{DD} -> R -> $ point $V_{12} -> C -> $ point $V_1 = 0$.** The charging process will go on till the voltage V_{12} is equal to the threshold voltage of the second NOR gate (inverter): $V_{12} = V_{IL}$. After that, the second NOT will invert and we obtain $V_{out} = 0$. Therefore, the output voltage will be high through the charging time from zero up to T which relates with the voltage V_{IL}. Figure 4.61 represents the one-shot circuit implemented using CMOS technology. To better understand this process, let us consider the behavior of the circuit in this figure. We will provide an analysis of this circuit using the timing diagram shown in Fig. 4.62.

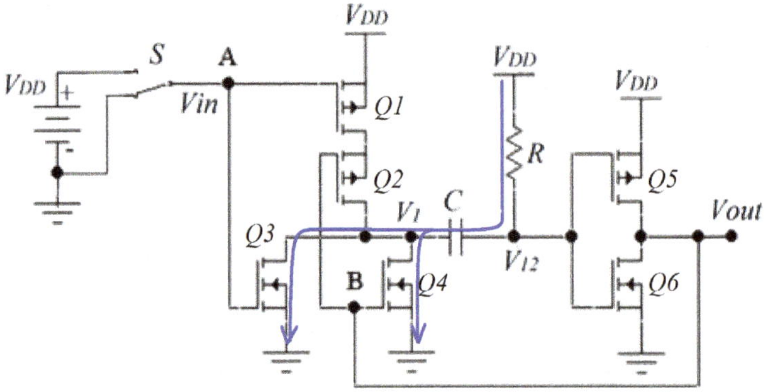

Fig. 4.61.   The monostable multivibrator circuit implemented by CMOS technology.

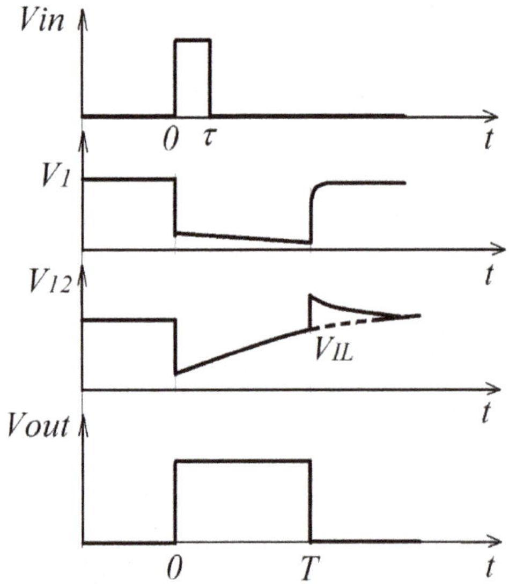

Fig. 4.62.   The timing diagram of the behavior of the monostable multivibrator.

In the initial state, the switch S connects the input point A to the ground. Therefore, the input voltage on the gates of transistors $Q1$ and $Q3$ is low: transistor $Q1$ is in the conducting state and transistor $Q3$ is closed. As this state (the initial state) is stable, a capacitor C should be discharged. The right side of the capacitor,

V_{12}, is connected to the supply V_{DD} through the resistor R. Thus, the left side of the capacitor also should be connected to the supply. It is only possible through two consequently connected transistors $Q1$ and $Q2$ when they both are in the conductive state. So, the point V_1 is connected to the supply. The high state of the point V_{12} causes the inverter $Q5$-$Q6$ to be in the zero-state. Therefore, the output signal V_{out} is low and it keeps the transistor $Q2$ in the open state and the transistor $Q4$ in the closed state.

At the time $t = 0$, the switch S connects the input voltage to the supply: the high voltage opens the transistor $Q3$ and closes the transistor $Q1$. The voltage V_1 falls to almost zero and this sharp drop passes through the capacitor as a high-frequency signal (δ-function). Now, the voltage V_{12} is low and the inverter $Q5$-$Q6$ causes a high logic state at the output point. This state is not stable. The capacitor begins to charge through the following electrical path: **Supply $V_{DD} -> R ->$ point $V_{12} -> C ->$ point $V_1 =$ transistor $Q3$ and transistor $Q4$ (in parallel) $->$ ground**. After the end of the input pulse t, the switch S returns to the initial state and closes the transistor $Q3$. The charging of the capacitor continues through the transistor $Q4$. Evidently, the input pulse duration should be more than the inverting time of the inverter $Q5$-$Q6$.

The charging current through a capacitor may be presented by the following equation:

$$i_C(t) = I_C(0)e^{-\frac{t}{\tau_1}}. \tag{4.30}$$

Where

$$I_C(0) = \frac{V_{DD}}{R + R_{on}}, \tau_1 = C(R + R_{on}) \quad \text{and}$$

$$R_{on} = \frac{1}{K_n(V_{DD} - V_T)} \text{(see Eq. (3.23))}.$$

The voltage on the capacitor V_{12} may be calculated using the equation

$$V_{12} = V_{DD} - i_C(t) \cdot R. \tag{4.31}$$

This voltage increases to reach the threshold value for the inverter $Q5$-$Q6$. After that, the inverter flips and the low voltage comes to point B, closes the transistor $Q4$ and opens the transistor $Q2$, thus beginning the discharge of the capacitor. We designate the full time required for the voltage V_{12} to increase up to the threshold voltage as T. If we insert the time $t = T$ in Eq. (4.30) and after arranging Eq. (4.31), we obtain the following:

$$V_{12} = V_{th} = V_{DD} - \frac{V_{DD}R}{R + R_{on}}e^{-\frac{T}{\tau_1}}.$$ (4.32)

This equation enables us to calculate the output pulse duration:

$$T = C(R + R_{on})ln\left[\frac{V_{DD}}{V_{DD} - V_{th}}\frac{R}{R + R_{on}}\right].$$ (4.33)

The threshold voltage in the inverter is the maximum input voltage keeping the logic "1" in the output of the inverter, so $V_{th} = V_{IL}$. However, sometimes the threshold voltage in such circuits is approximately equal to $0.5V_{DD}$. In this case, if the value R_{on} is low, in other words, if $R_{on} \ll R$, Eq. (4.33) will decrease to the reduced form:

$$T = CRln2 = 0.69CR.$$ (4.34)

4.4.4.2. *Astable multivibrator*

An astable multivibrator begins to generate the output meander-like signal immediately after the connection to the power supply. Figure 4.63 represents the simplest circuit implementing the astable multivibrator behavior.

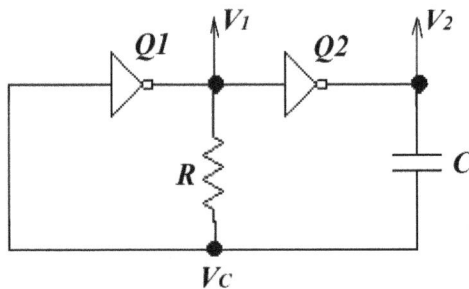

Fig. 4.63.   The simplest astable multivibrator based on two inverters.

The physical meaning of an astable multivibrator circuit is in the energy pumping from one place to another using a capacitor as the energy store. Here, we are looking at the swapping process of two inverters switching with the continuous pumping from the supply. In this circuit, the charging-discharging process continues during the time that the inverters are supplied from the external energy supply. The output voltages V_1 and V_2 cause the meander-like voltages to enter the opposite phases. The inverters $Q1$ and $Q2$ may be built using all the types of technologies which we have considered. To better understand the astable multivibrator behavior, let us consider the circuit constructed with CMOS technology as shown in Fig. 4.64.

The circuit presented in Fig. 4.64 is accompanied by the timing diagram shown in Fig. 4.65.

As known, the output voltages of the CMOS inverter are zero and V_{DD}. Therefore, in the case of CMOS inverters, we do not need to take into account the resistances of the open transistors included in the inverter's circuit. Evidently, when another technology is applied, these resistances will be a part of the calculations defining the oscillation period. Also, the duty cycle in these oscillations is equal to 50%, as the charge and the discharge currents flow through the same way.

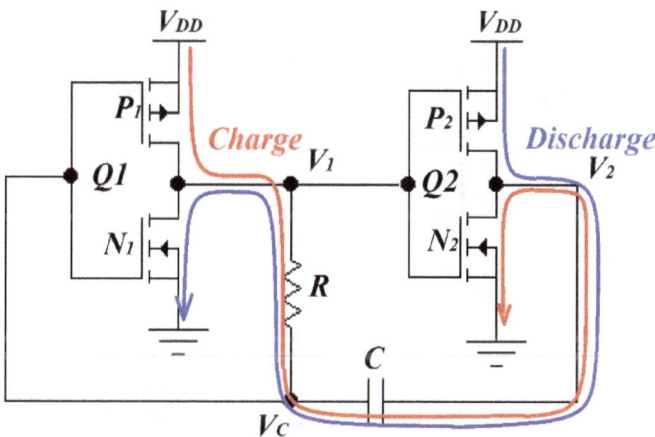

Fig. 4.64.   The astable multivibrator behavior.

Let us begin our analysis of this circuit at the point when $V_1 = V_{DD}$ and $V_2 = 0$. In this state, the voltage V_C increases exponentially through the circuit **V_{DD} -> point V_1 -> R -> point V_C -> C -> point V_2 -> zero.** This charging process occurs until the voltage V_C is not equal to the threshold voltage V_{th}. At the time $t = T/2$, the voltage V_C becomes equal to V_{th}, the inverter $Q1$ switches to the state "0" and the inverter $Q2$ switches to the state "1" or V_{DD}. This jump from zero to V_{DD} passes through the capacitor and is added to the voltage V_{th}. Thus, the voltage V_C will be equal to $(V_{th} + V_{DD})$, which begins to discharge through the resistor R at the point $V_1 = 0$. The duration of the discharge process is equal to $T/2$; we need this time to decrease the voltage V_C to V_{th}. After that, both inverters switch once more and the voltage V_C falls to the value $-V_{th}$. Here, the capacitor begins to charge once more and everything that has occured repeats itself. This multivibrator keeps the oscillation frequency, which only changes due to external temperature changes.

To calculate the oscillation frequency, we should take into account that the output voltage of the CMOS inverter is equal to zero or V_{DD} and the threshold voltage is equal to V_{IH} (see Eq. (3.78)) or V_{IL} (see

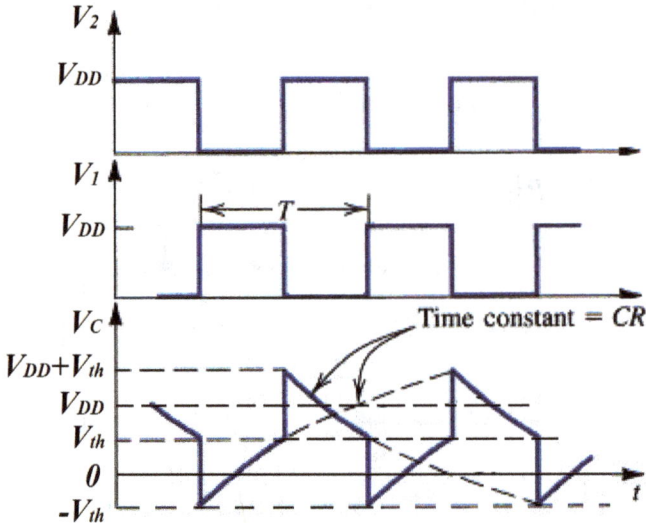

Fig. 4.65. Timing diagram describing behavior of the charge-discharge process.

Eq. (3.76)). Sometimes, this threshold voltage may be approximately given as $V_{th} = \frac{V_{DD}}{2}$. In this way, the voltage at point V_1 is equal to the constant value V_{DD} or 0 always. So, for the charging process marked in red, we can write the following equation:

$$V_C = V_{DD} - i_c(R + R_{on}) \tag{4.35}$$

where

$$i_c = I_0 e^{-\frac{t}{\tau}} = \frac{V_{DD}}{R + R_{on}} e^{-\frac{t}{\tau}} \ and \ \tau = C(R + R_{on}). \tag{4.36}$$

When the voltage V_C reaches V_{th}, Eq. (4.35) transforms into the following:

$$V_{DD} - V_{th} = \frac{V_{DD}(R + R_{on})}{R + R_{on}} e^{-\frac{T}{C(R+R_{on})}} \tag{4.37}$$

that after rearrangement produces the following:

$$T = -C(R + R_{on}) ln \frac{V_{DD} - V_{th}}{V_{DD}}. \tag{4.38}$$

Evidently, in the case $V_{th} = V_{DD}/2$ and $R_{on} \ll R$, this equation reduces to the form $T = -CRln2 = 0.69 \ CR$.

There are many different astable multivibrator circuits based on the described system. For example, the circuit presented in Fig. 4.66 enables us to build a multivibrator with a controlled duty cycle. Here, the duty cycle is defined by the relation between two resistors R_1 and R_2.

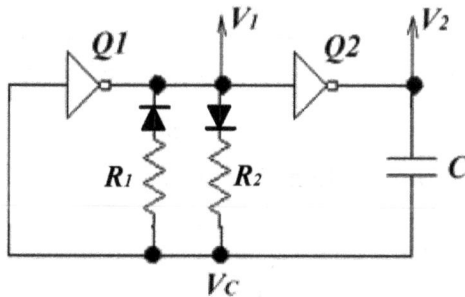

Fig. 4.66.   The astable multivibrator with different charge and discharge paths.

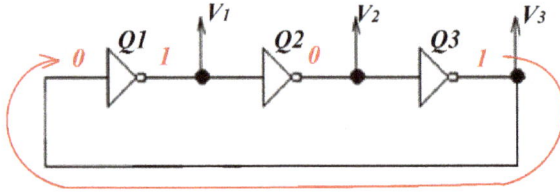

Fig. 4.67.   The astable multivibrator consisting of an odd number of inverters.

One other simple type of astable multivibrator is based on the number of consequently connected inverters. As known, two inverters connected in the ring according to Fig. 4.32, appear the latch circuits which remain the same state through all the time of working. Moreover, an even number of inverters connected in a ring behaves in the same way. However, a circuit consisting of an odd number of inverters connected in a ring presents another picture. Figure 4.67 represents the system with a minimum of connected devices.

As shown here, this circuit cannot remain in the same state. Each inverter requires a time delay to transfer the input voltage to the output. Therefore, the circuit's behavior may be understood as the result of the consequent but not instantaneous transfer of the input logic function to the output logic function. The output of each inverter is used as the input for the next one. The last output is fed back to the first inverter. Because of the delay time of each stage the whole circuit spontaneously starts oscillating at a certain frequency. This logic function represents the superposition of three parallel square waves related to three connected inverters. In the case of another odd number of inverters, the number of such wave functions will be equal to the number of inverters. The period of oscillation is proportional to the number of connected inverters:

$$T = 2nt_{PHL} \qquad (4.39)$$

where t_{PHL} is the propagation time delay for the individual inverter (see Eq. (3.90)).

The behavior of the ring oscillator may be characterized by the timing diagrams presented in Fig. 4.68.

These ring oscillators are very simple and stable multivibrators and can work in the digital circuits as timers. One requirement

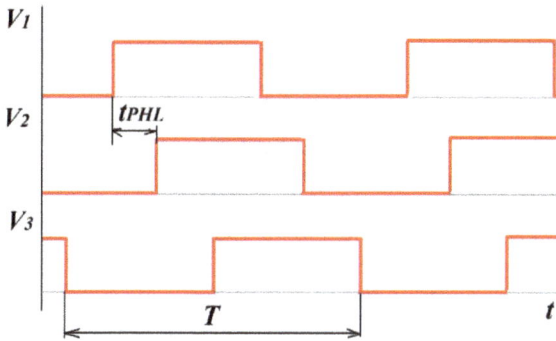

Fig. 4.68.   The timing diagrams of the astable ring-oscillator.

for the frequency to remain stable is the constant environment temperature.

4.5.   Problems

4.5.1 Figure 4.69 presents a logic circuit.

a. Identify the logic function implemented by the presented circuit.
b. Using CMOS technology, design a circuit implementing the same function.
c. Using PTL-CMOS technology, design a circuit implementing the same function.
d. It is known that the parameters of the basic (equivalent) CMOS inverter are as follows: $L = 2$ μm, $(W/L)_n = 3/2$, $\mu_n/\mu_p = 2.5$.

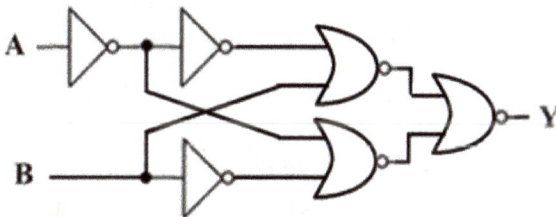

Fig. 4.69.   Logic circuit.

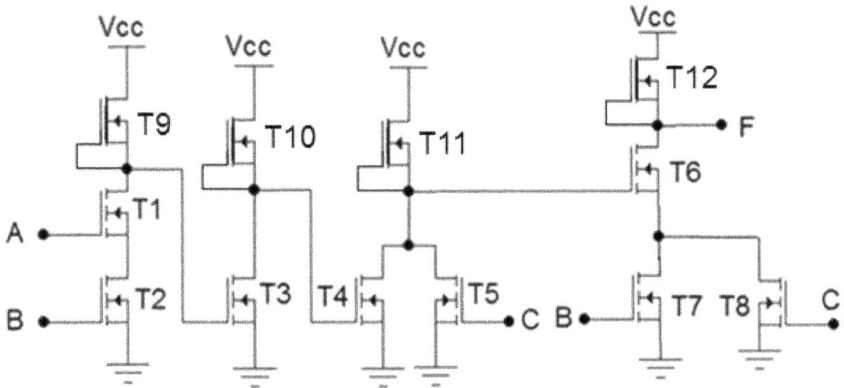

Fig. 4.70.   Logic circuit.

Calculate the full surface area occupied by the circuits from parts b and c.

4.5.2 Figure 4.70 presents a logic circuit.

    a. Identify the logic function implemented by the presented circuit.

    b. Using CMOS technology, design a circuit implementing the same function.

    c. Using PTL-CMOS technology, design a circuit implementing the same function.

    d. It is known that parameters of the basic (equivalent) CMOS inverter are as follows: $L = 2$ μm, $(W/L)_n = 3/2$, $m_n/m_p = 2.5$.

    e. It is known that Vcc $= 5$V, $K'_n = 5\mu$A/V, $|V_{TNL}| = V_{TND} = 1$V, and the load capacitance is equal to 10 pF; calculate the t_{PHL} and t_{PLH} for the circuit from part b.

Calculate the full surface area occupied by the circuits from parts b and c.

4.5.3 Figure 4.71 presents a logic circuit.

    a. Identify the logic function implemented by the presented circuit.

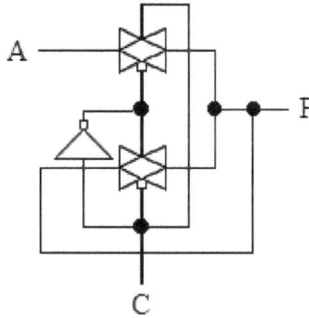

Fig. 4.71. Logic circuit.

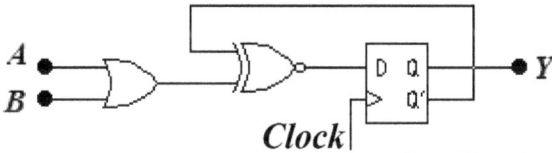

Fig. 4.72. Logic circuit.

    b. Using CMOS technology, design a circuit implementing the same function.

4.5.4 Figure 4.72 presents a logic circuit.

    a. Identify the logic function implemented by the presented circuit.

    b. Using CMOS technology, design a circuit implementing the same function.

    c. Using PTL-CMOS technology, design a circuit implementing the same function.

4.5.5 Figure 4.73 presents a digital multivibrator circuit.

It is known that the supply voltage is 1.8 V, the work frequency is 100 MHz, and the current produced by the oscillator through one period is 200 mA.

Calculate the dynamical power of each inverter in the circuit, the input capacity of each inverter and the propagation time delay of each inverter.

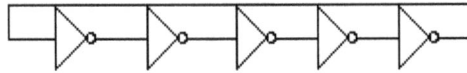

Fig. 4.73.   Multivibrator circuit.

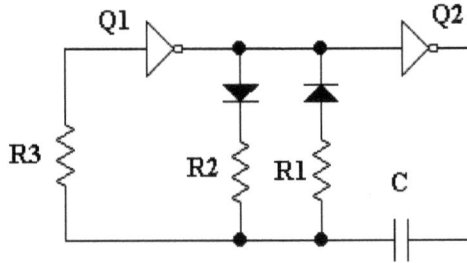

Fig. 4.74.   Non-stable multivibrator.

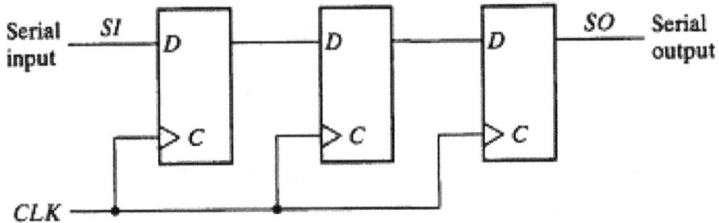

Fig. 4.75.   Digital synchronous circuit.

4.5.6 Figure 4.74 presents a non-stable oscillating circuit using CMOS inverters.

The circuit has the following parameters: $(W/L)_n = 2$, $K' = 25 \ \mu A/V^2$, $V_{Tn} = |V_{Tp}| = 1$ V, $V_{DD} = 5$ V, $C = 1 \ \mu F$, $V_{BE,on} = 0.7$ V, $\mu_n/\mu_p = 2.5$.

Assuming that the threshold voltage of the inverters is equal to $\frac{V_{DD}}{2}$, calculate the resistors R1 and R2, provided that the square wave with the period $T = 50$ msec and the duty cycle DC = 20%.

4.5.7 Figure 4.75 presents a digital synchronous circuit.

     a. Identify the logic function implemented by the presented circuit.

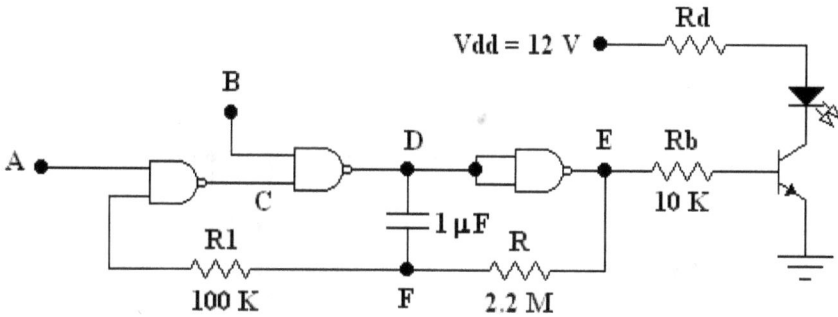

Fig. 4.76. Digital logic circuit.

b. Design a circuit implementing the same function using inverters of the CMOS type and switches of the PTL-CMOS type.

c. It is known that the transistor's length equals to $L = 2\mu m$, $\mu_n = 1450$ cm^2/V×s, $\mu_n/\mu_p = 3$, $(W/L)_n = 2$, $K'_n = 30$ $\mu A/V^2$, and the equivalent input capacity of the CMOS inverters equal to 10 fF. Calculate the propagation time of the rising edge of the pulse from the input to the output of the circuit.

4.5.8 Figure 4.76 presents a digital logic circuit.

All the logic gates in the circuit are designed using CMOS technology. It is known that the threshold voltage of the CMOS gates is $V_{Th} = 0.5 Vdd = 6$ V, that the LED produces the current $I_{LED} = 15$ mA under the applied voltage of $V_{LED} = 1.5$ V, and that the parameters of the transistor are as follows: the magnification coefficient $b = 30$ and the voltage of the transistor opening $V_{BE} = 0.6$ V.

a. Identify the status of the LED (Light/Darkness) for the following states of the inputs:

A	B	LED
Low	Low	
Low	High	
High	Low	

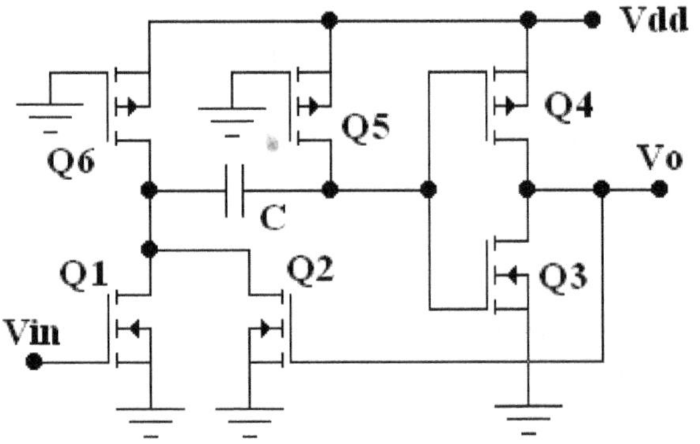

Fig. 4.77.   Digital logic circuit.

b. What happens if $A = B =$ High?

c. Calculate the resistance of the resistor Rd.

d. According to the calculated times and voltages, draw the waveforms at points D, E, and F.

4.5.9 Figure 4.77 presents a digital logic circuit. It is known that the supply voltage is equal to 3.3 V, the minimal transistor length $L_{min} = 0.8\mu m$, $V_{TN} = -V_{TP} = 0.6$ V, $k'_p = 40$ $\mu A/V^2$, $k'_n = 100\mu A/V^2$, $[W/L]_{min} = 1.2/0.8$, Cox $= 1.8$ fF/μm^2, $C = 10$ pF. All transistors of the NMOS type have the same $[W/L]_{min} = 1.2/0.8$ and L_{min}. Transistor Q6 has minimal dimensions, transistors Q3 and Q4 are complementary, and transistor Q5 has minimal width and its length is more than the minimum by 10 times.

a. Calculate the width of the output pulse.

b. Calculate the minimal width of the input pulse that will cause changes in the circuit.

c. Calculate the maximum current flowing through the capacitor immediately after the circuit switches to the high output state.

4.5.10 Figure 4.78 presents a digital logic circuit.

a. Write the logic function of the presented circuit.

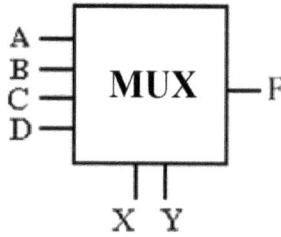

Fig. 4.78. Digital multiplexer circuit.

b. Using CMOS technology, design a circuit implementing the same function.
c. Using PTL-CMOS technology, design a circuit implementing the same function.
d. It is known that the parameters of the basic (equivalent) CMOS inverter are as follows:

$$L = 2\ \mu m,\ (W/L)_n = 2,\ m_n/m_p = 2.5.$$

Calculate the full surface area occupied by the circuits from parts b and c.

Chapter 5

Semiconductor Memory Architecture

5.1. Semiconductor memory architecture

Digital circuits are the elemental base of digital computers. Digital computers appeared in the middle of the last century. The first computers were created using vacuum tubes (electronic lamps). Figure 5.1 presents such electronic lamps: diodes, triodes, tetrodes, etc.

A computer comprises various modules. One of them is a memory module. This module is devoted to the storage, writing, and reading

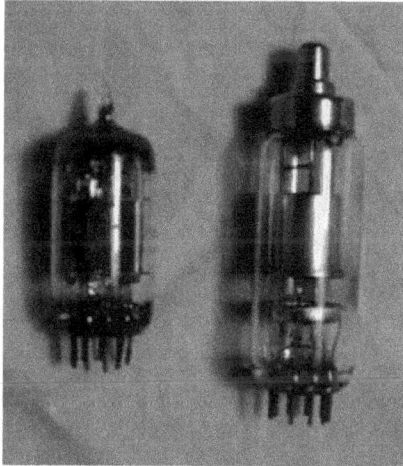

Fig. 5.1. Electronic lamps applicable in digital computers.

of various types of information: constants, data, and instructions for the execution of calculations. We have studied already such logic gates as flip-flops. Each flip-flop can store one bit of information. Several flip-flops aggregated together are called a register, which can store a series of information bits. In this way, the memory of a computer may be presented as a collection of such registers. The length of the register or the number of flip-flops constituting the register can be used to store a computer word which can represent various types of information. A piece of information that consists of a suitable computer word written in flip-flops can be read or changed simultaneously. The calculation ability of the computers was very limited based on the vacuum tube's technology. Their main restrictions are big dimensions, high power consumption, and low execution rate. For example, after one hour of calculation, the computer had to be disconnected from the power supply due to overheating. Figure 5.2 presents one of the first computers, "ENIAC", which was used in 1945. Such computers required large areas and a lot of qualified staff.

Due to the high demand for computers, progress in this area was very rapid. A magnetic memory based on ferrite rings was invented in the middle of the last century and was the predominant form of memory applied in industry from 1955 up to approximately 1975. Figure 5.3 represents a memory cell based on the ferrite ring.

Figure 5.3(a) presents the operational principle of the memory unit and Fig. 5.3(b) illustrates a magnetic memory element (ME). Here, the memory element has input (write) and output (read) possibilities. To choose the defined memory element we must select it. In this way, we obtain an element which can be inserted into a large net consisting of many such elements. Thus, memory may be presented as a net with rows and columns. Figure 5.4 illustrates this memory net. There are memory elements at all the intersection points.

The ferrite ring, which has a rectangular ferromagnetic hysteresis loop, is the main element storing information. The principle of operation of the magnetic element is shown in Fig. 5.5. A ferromagnetic element-ring has two stable states defined by the direction of the current passing through it. The magnetic field stored in the

Fig. 5.2.   One of the first digital computers.

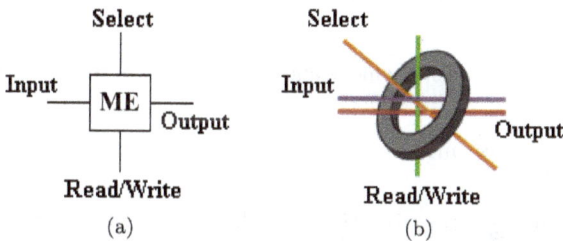

Fig. 5.3.   The principal scheme of the magnetic memory cell.

ferromagnetic ring may have two different values: $+1$ and -1. This field continues to be retained in the ring even after the current cutoff. Due to these properties, the magnetic memory is a non-volatile memory.

To change the state of the memory element, we need to change the direction of the current passing through the wire and the ring.

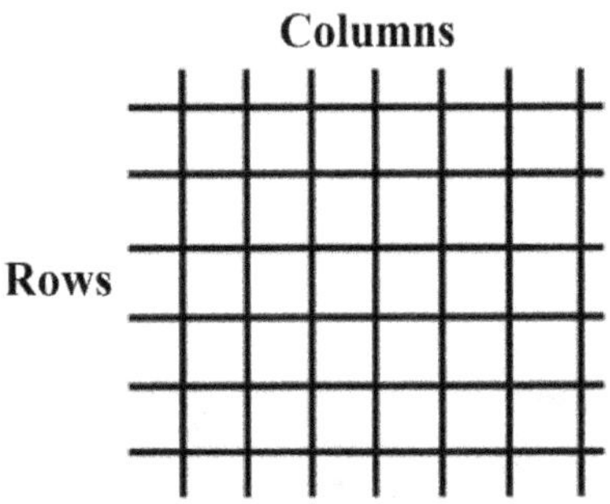

Fig. 5.4. A principal scheme of the memory net.

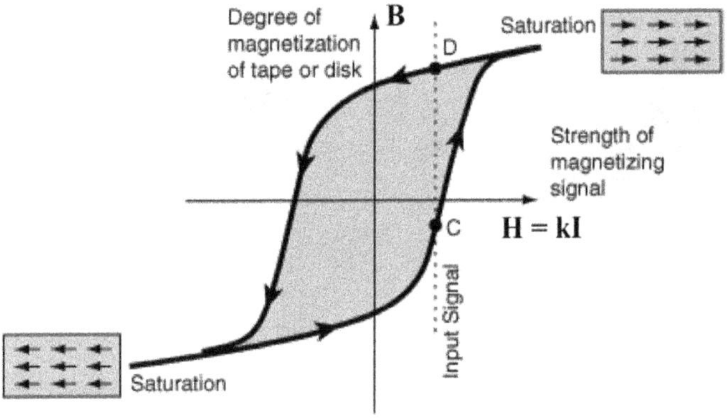

Fig. 5.5. The magnetic memory and the principle of information storage.

A memory module built on a base of ferromagnetic rings is shown in Fig. 5.6.

The following step in the development of digital computers occurred after the invention of semiconductor bipolar transistors (BJT). The basic logic family which was used for memory construction with BJT was the TTL logic family. The same net principle was applied to the arrangement of this form of memory. Such a net enables us to get information from memory or to write information to each memory cell at the same time independently of the place of the cell in the net. Such a type of memory was called random-access memory (RAM). Figure 5.7 represents a memory element based on the SR flip-flop. Once more, as in the magnetic memory, the core element

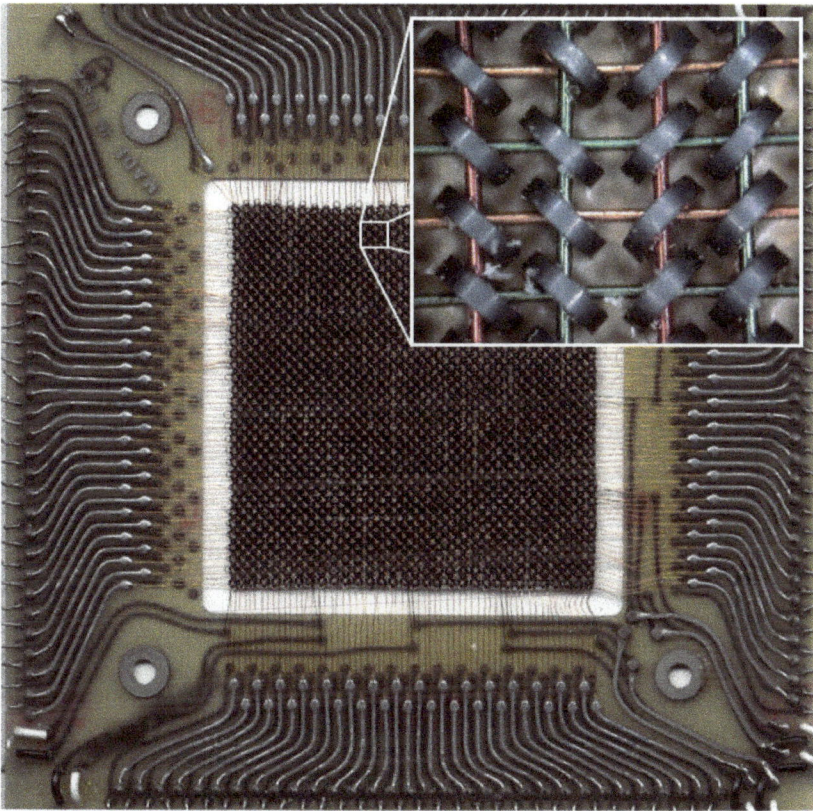

Fig. 5.6. A memory module based on the ferromagnetic rings.

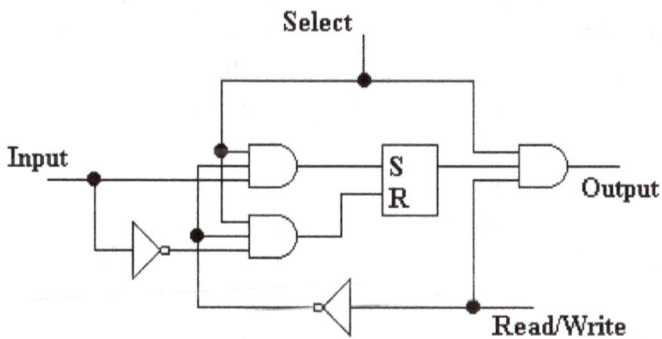

Fig. 5.7. A logic scheme of the memory element used in the RAM.

of the memory cell is a gate enabling two stable states. Here, the SR flip-flop plays the role of a digital element storing the information bit. Such a scheme may be easily used in building the memory net.

The logic circuit presented in Fig. 5.7 works as follows. Each signal connected with the memory element may be in the two different logic states "1" or "0". So, two opposite actions, read and write cannot occur simultaneously. We know that the flip-flop is a logic circuit that is dependent on the time, therefore this element always stores a suitable logic state, "1" or "0". To read this information, we need to take a Read/Write signal equal to "1". This signal will enable the information to move to the output and disable the charge of the memory cell. If the cell is selected, the information goes to the output node and will be read. A Read/Write signal equal to "0" disables our ability to read the information stored in the cell. At the same time, it enables "Set" in the flip-flop so that it is in the required state for writing. Such a logic scheme may be easily embedded into the memory net.

The first MOSFET transistors were patented in 1959 by D. Kahng and M. Atalla from "Bell Labs". This invention had ushered in a novel era in electronics. The miniaturization of electronic devices has become the main direction in the development of electronics. Figure 5.8 represents two commercial integral circuits: the logic gate created in 1960 and the microprocessor Intel 486 created in 1989. The first scheme consists of four transistors and two resistors. The second scheme consists of approximately 6 million transistors arranged on a surface with an area less than 1 cm^2. This device consumes low power, has small dimensions, and can produce approximately $40 \cdot 10^6$ instructions per second. The devices of the present day contain hundreds of millions of transistors in the same area and work with a frequency of several gigahertz.

Now, we begin to consider the memory organization based on the semiconductor devices, specifically the NMOS and CMOS technologies. Figure 5.9 illustrates various types of semiconductor memory and the relation between the different types.

Semiconductor memory may be divided into three big groups, each one of them used to solve different problems. The first group

Fairchild, Gordon Moore, 1960

Fig. 5.8. Two integral circuits produced with an interval of about 30 years between them: the "Fairchild" logic gate consisting of four transistors and two resistors and the Intel-486 microprocessor.

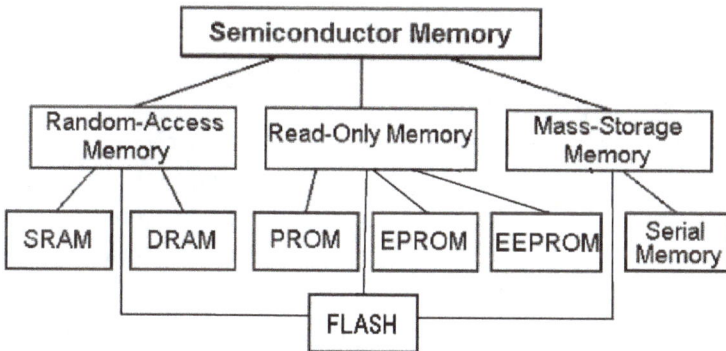

Fig. 5.9. A structure organization of the semiconductor memory.

is the random-access memory or RAM which enables us to store instructions, data, and various constants. The main property of RAM is our ability to access any piece of information stored in memory in the same amount of time independent of its location in memory. Due to the general principle of information storage, each information bit may be stored in two ways. The first method of storage is the application of latch schemes; the second one is the application of a capacitor for this goal. Consequently, all RAM systems are divided into systems that use static random-access memory (SRAM) and systems that use dynamic random-access memory or DRAM. SRAM is built on the base of latch-schemes and capable of storing information for a long time, up to the supply cutoff. DRAM requires a periodic refresh of the stored information due to leakages in the semiconductor capacitors.

The second big group of memories is read-only memory (ROM). ROM is used to store significant information which should be stored without any changes throughout the lifetime of the device. The main operating system of a computer and very important information are usually written in ROM, for example a game which may only be changed with the ROM exchange. The construction of ROM has changed with the development of novel transistor technologies. The first ROMs were programmed with fuses firing in memory; these devices were called PROM. After that, the electrically programmed read-only memory devices (EPROM) were designed. We will discuss the construction of EPROM later. With the goal of changing information in the ROM, erasable devices called EEPROM were designed and applied in industry.

The third group of memory devices is used for the long-time storage of information. At the beginning, carton cards with perforated writing were used for information storage. They were replaced by magnetic storage. Figure 5.10 shows a roll of magnetic tape and a read/write arrangement applicable for computers.

Magnetic storage was realized using a polymer tape coated by a specific powder containing ferromagnetic material, for example permalloy. Also, for information storage, polymer disks with a magnetic coating were used. These disks, called "floppy disks" due to their flexibility, and disks of different sizes were able to store different volumes of information. Figure 5.11 presents the various floppy disks and their drivers — the devices able to read and write information stored in the floppy disks and exchange the information with a computer. The main deficiency of these storage systems is the limited volume of information which may be stored.

Fig. 5.10. A roll of magnetic storage tape and a read/write device.

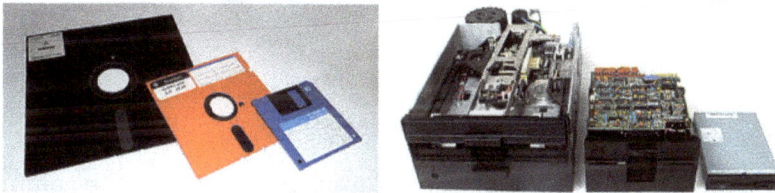

Fig. 5.11. 8-inch, $5\frac{1}{4}$ -inch, and $3\frac{1}{2}$ -inch floppy disks and the suitable drives.

One can see further development of the floppy disk systems in the disk called "ZIP". 100 MB of information may be stored on such a system. Figure 5.12 shows the "ZIP" floppy disk and a suitable driver.

Unfortunately, these systems were short-lived. Approximately in 2007 they were increasingly being replaced by other forms of digital storage. The compact disks were built and introduced into the market due to their capacity of ~650 MB. Figure 5.13 shows the CD-ROM disk and a driver. The compact disk (CD) is a digital optical disc data storage format released in 1982 and co-developed by Philips and Sony. The format was originally developed to store and play only sound recordings but was later adapted for the storage of data (CD-ROM).

Fig. 5.12.   100 MB "ZIP" floppy disks and the suitable driver.

Fig. 5.13.   Compact disk and the suitable driver.

Figure 5.14 illustrates the structure and elements of the compact disk. A CD is made from 1.2 millimeters (0.047 in) thick, polycarbonate plastic and weighs 15–20 grams. From the center outward, its components are: the center spindle hole (15 mm), the first-transition area (clamping ring), the clamping area (stacking ring), the second-transition area (mirror band), the program (data) area, and the rim.

A. A polycarbonate disc layer has the data encoded by using bumps.
B. A shiny layer reflects the laser.
C. A layer of lacquer protects the shiny layer.
D. Artwork is screen printed on the top of the disc.
E. A laser beam reads the CD and is reflected back to a sensor, which converts it into electronic data.

The inner program area occupies a radius from 25 to 58 mm. A thin layer of aluminum or, more rarely, gold is applied to the

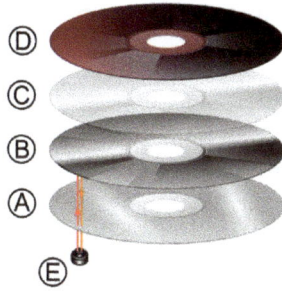

Fig. 5.14. A diagram of the CD layers.

surface, making it reflective. The metal is protected by a film of lacquer normally spin coated directly on the reflective layer. The label is printed on the lacquer layer, usually by screen printing or offset printing. CD data is represented as tiny indentations known as "pits," encoded in a spiral track molded into the top of the polycarbonate layer. The areas between pits are known as "lands." Each pit is approximately 100 nm deep by 500 nm wide and varies from 850 nm to 3.5 μm in length. The distance between the tracks, the pitch, is 1.6 μm. The program area is 86.05 cm^2 and the length of the recordable spiral is $(86.05 \text{ cm}^2/1.6 \, \mu\text{m}) = 5.38$ km. With a scanning speed of 1.2 m/s, the playing time is 74 minutes. 650 MB of data can be stored on a CD-ROM. Information may be recorded on the CD in various formats. These types are presented in Fig. 5.15.

The difference is in the dimensions of the pits and in the wavelength of the read/write device. Memory storage systems based on optical compact disks were very common until approximately 2006 when the flash memory systems began to enter the market. Figure 5.16 represents a side view of a transistor with a floating gate.

As shown in the figure, a flash transistor has two gates with different functions. The first, a floating gate, is in the environment of a dielectric insolating layer which cannot transit charged carriers in normal conditions. The main function of this gate is to store the negative charge and prevent the building of an inversion layer in the transistor's body. The second, a control gate, is connected with other elements of the circuit and causes the switching of

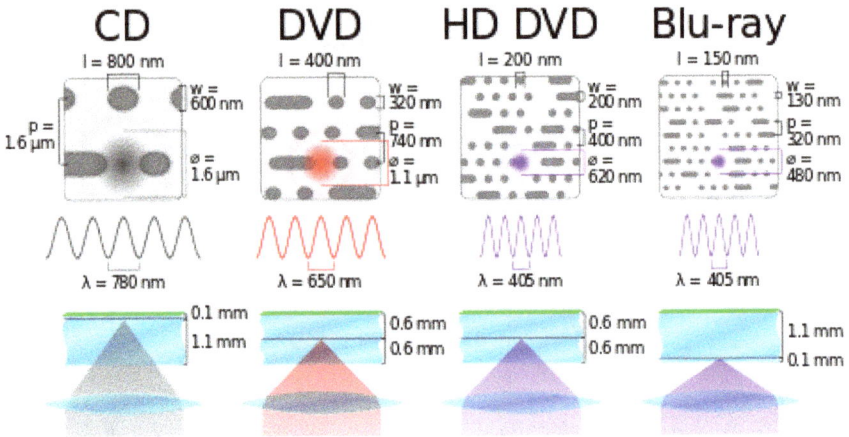

Fig. 5.15.   Comparison of various storage media.

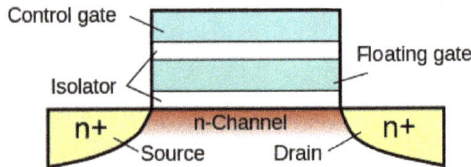

Fig. 5.16.   The cross-section of a floating-gate transistor.

the transistor to the programmed state and back. A programming (writing) voltage is more than the usual control voltage representing a logic "1". Figure 5.17 illustrates the programming-erasing processes.

To program the transistor, a high voltage is applied between the source and the drain. This voltage leads to the appearance of very high-speed electrons which jump to the control gate that is also under high voltage. A number of these electrons enter into the floating gate and prepare an equipotential surface there which prevents further electron jumping. Now the floating gate is charged and the conductive layer cannot be built between source and drain. To erase this charge, the reverse voltage should be applied between a control gate and the drain. Figure 5.18 presents (a) a flash-transistor used in memory applications and (b) the internals of a typical USB flash drive.

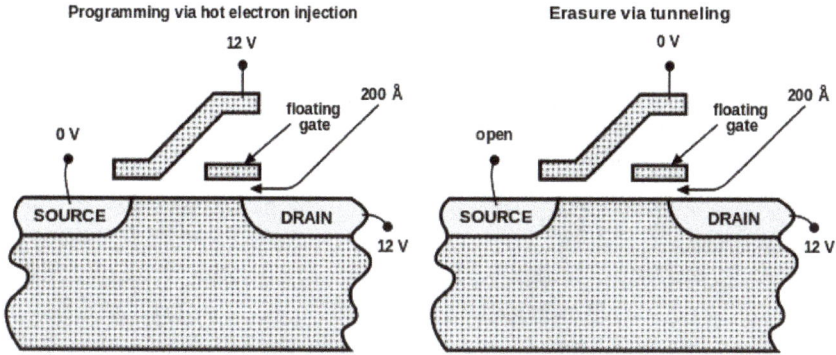

Fig. 5.17. Programming a NOR memory cell (setting it to logical 0), via hot-electron injection and erasing a NOR memory cell (setting it to logical 1), via quantum tunneling.

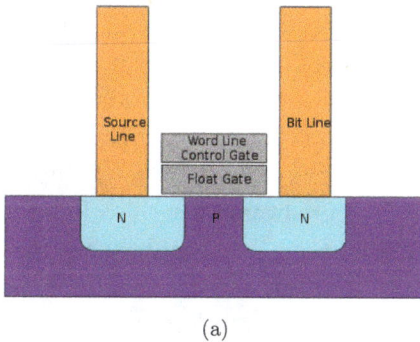 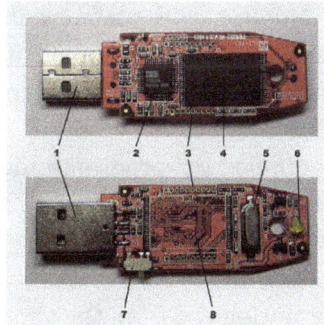

(a) (b)

Fig. 5.18. (a) The principal construction of a flash memory cell and (b) the internal view of a typical USB flash drive.

Here, Fig. 5.18(a) illustrates the principal arrangement of a one-transistor memory cell. We can see the word line control gate which connected with the word-line in the memory net. Also, the bit-line electrode which connected with the transistor's drain is shown. Using the control circuit, a specific voltage is applied to suitable electrodes to read or write a piece of information in the chosen memory cell. The second part of the figure, part (b), shows a typical internal view of the flash memory device. It consists of the following elements: 1 — USB Standard-A "male" plug, 2 — USB mass storage controller,

3 — test point, 4 — flash memory chip, 5 — crystal oscillator, 6 —
LED (optional), 7 — write-protect switch (optional), and 8 — space
for a second flash memory chip.

Now, flash memory is widely used in the world. The development
of the flash memory elements of enlarged capacity led to these
elements being applied in all types of memory: RAM, ROM, and
mass-storage elements. Flash memory was used in the last series of
computers instead of hard drives; there are mobile external flash
memory discs with the capacity of terabits. So, the last generation
of computers includes memory devices based on flash memory.

We have briefly considered various types of memory devices based
on different technologies. However, the main common principle of
their design is the creation of the memory devices according to the
simple net shown in Fig. 5.4. Figure 5.19 presents the net structure
of random-access memory in more detail. As shown, a memory is
presented using a number of nets connected in parallel. A number
of nets connected in parallel is equal to the "word" used in this
digital system. The "word" applied in the system is a number of
bits applied for programming instruction, data, and constants. The
word may be equal to 16, 32, 64, etc. bits according to the applied
system architecture and hardware oriented language, also known as
an assembler. The system, shown in Fig. 5.19, enables us to get
information out from the memory at the same rate independent of the
location of this information. In other words, random access to each
memory cell is provided in the presented system. The system consists
of memory nets and environment circuits providing access to the cells:
a row decoder, a column decoder, a sense amplifier series and an
input-output (I/O) terminal enabling connection with the memory.

The memory system works as follows: each instruction consists
of an operation code and an address. The address consists of a
row code and a column code which is transferred to the row and
column decoders. Each decoder is a combinational circuit, therefore
the generation of a row's number after the code is received takes no
time in the ideal case. In reality, the transfer of a code through the
decoder and the generation of the row's number requires the same
time independently of the real address of the memory cell. After

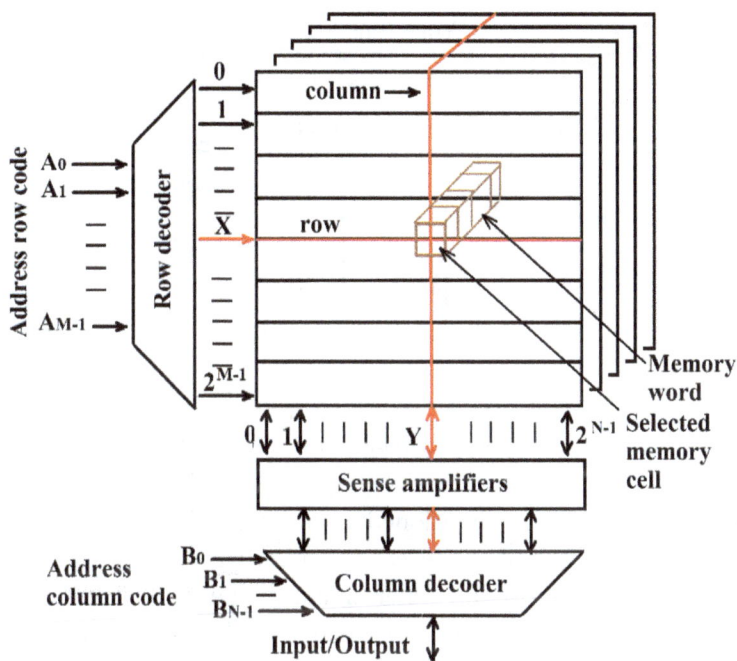

Fig. 5.19. The principal scheme of a memory device.

choosing the row, the column will be chosen and information from the cell will be transferred to the I/O terminal or written to the cell from the I/O terminal. The sequence of all memory operations is controlled using a timer and suitable control signals.

When a memory cell is chosen, all the memory cells that are in the parallel memory nets will also be chosen and a full word will be sent on the I/O terminal. The row decoder can address 2^M rows if M designates length of the code (in bits) in accordance with the selected row. The column decoder addresses 2^N columns. Therefore, the presented memory net can address 2^{M+N} memory words. Thus, a memory net with $M = N = 10$ consists of 10^6 memory cells.

5.2. Random access memory cells (SRAM)

Let us consider the SRAM memory cell in more detail. Figure 5.20 presents an SRAM memory cell. This memory cell represents a

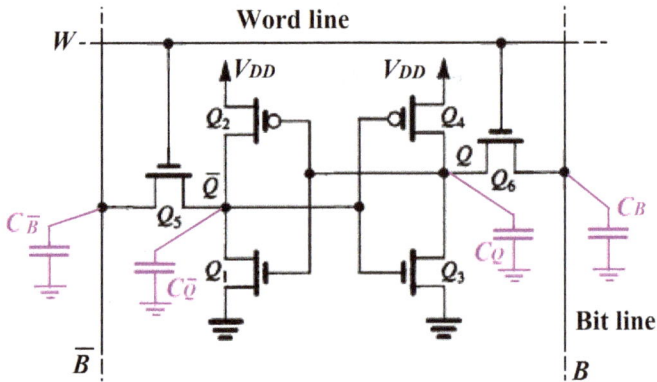

Fig. 5.20. A principal circuit of the SRAM memory cell.

simplified S-R flip-flop mentioned above (see Fig. 4.42). Now the input S and R electrodes are connected with suitable bit-lines: B and $\bar{B}$. Also, instead of the timer signal $\emptyset$, the word line related to some row's number W is applied to control the pass transistors $Q5$ and $Q6$. In other words, two CMOS inverters are connected in series to the ring so that they form a latch. The inputs of these inverters are connected with the bit lines through the pass-transistors $Q5$ and $Q6$. The input capacity of the inverter $Q_1 - Q_2$ is presented by the virtual capacitor C_Q. The input capacity of the inverter $Q_3 - Q_4$ is presented by the virtual capacitor $C_{\bar{Q}}$. Also, the bit-lines have suitable virtual capacities as these lines are not connected with the supply or ground. The information bit which should be stored in this memory cell represents a charge presented in the capacitor C_Q. It may be equal to V_{DD} for the logic "1" or zero for the logic "0".

All operations involving memory in digital systems such as computers occur periodically according to suitable control signals. Before the operations of writing or reading from the memory cell occur, the capacitors of bit-lines should be charged up to the voltage $V_{DD}/2$. This operation (precharging) is performed using a specific electrical circuit (equalizator) which simultaneously charges all bit-lines in the memory net for the defined voltage. The equalizator principal electric scheme is shown in Fig. 5.21. This circuit consists of three NMOS pass-transistors connected with the supply, which

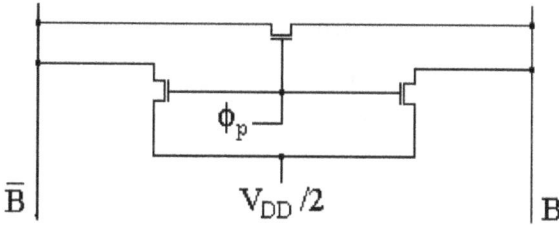

Fig. 5.21. A principal circuit of the equalizator.

produces a voltage $V_{DD}/2$. This voltage will be transferred to the bit-lines when the transistors are switched on by the periodical signal ϕ_p (precharge). The time duration of the precharge signal should be enough to allow the charging of bit-lines capacitors to the defined voltage. During the precharge, all the other functions of the memory are forbidden.

Let us consider a reading operation from the memory cell with the assumption that $Q = $ "1". In this state, the complement $\bar{Q} = $ "0". As was mentioned above, before reading all bit-lines in the memory are charged up to $V_{DD}/2$. Now a decoder system chooses a suitable word-line and assigns $W = $ "1" $= V_{DD}$. This time, both transistors Q_5 and Q_6 begin conducting and the currents begin to flow through them. As the voltage stored in the capacitor $C_{\bar{B}}$ is more than the voltage in the capacitor $C_{\bar{Q}}$ and the voltage stored in the capacitor C_Q is more than the voltage in the capacitor C_B, both currents will flow in the right direction and the voltage difference $\Delta V > 0$ will appear between bit-lines as shown in Fig. 5.22.

The voltage difference ΔV between bit-lines comes to the sense amplifier which is connected to the same bit-lines. This sense amplifier is presented in Fig. 5.23. It works as follows: at the beginning of the reading process, the sense amplifier is in the steady state and both the inverters that are connected in the ring shape have the same voltage in the inputs as in the outputs. After choosing the word-line $W = 1$, the sense signals $\phi_s = V_{DD}$ and $\bar{\phi}_s = 0$ simultaneously turn on the transistors Q_5 and Q_6 and the sense amplifier begins to change their state according to the voltage difference ΔV sign. In our case, the voltage difference is positive, therefore the inverter $Q_1 - Q_2$

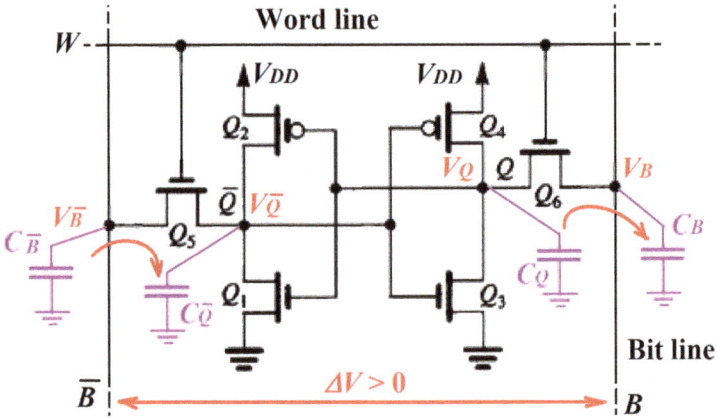

Fig. 5.22.   The reading process begins to flow.

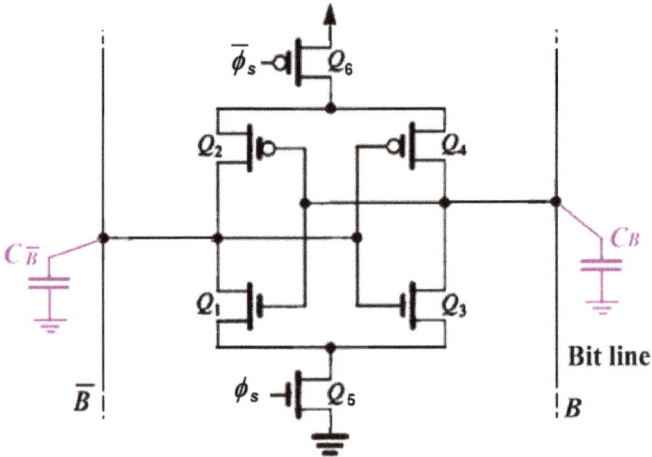

Fig. 5.23.   The sense amplifier's principal electrical circuit.

begins to go to the low state such that the transistor Q_1 will open a little bit and the transistor Q_2 will close by the same value. This change will act on the inverter $Q_3 - Q_4$ such that transistor Q_3 will close a little bit and the transistor Q_4 will open by the same value. In this way, using the described positive feedback, the inverter $Q_1 - Q_2$ will reach the logic "0" state and the inverter $Q_3 - Q_4$ will reach the logic "1". This whole process occurs very quickly and the bit-lines

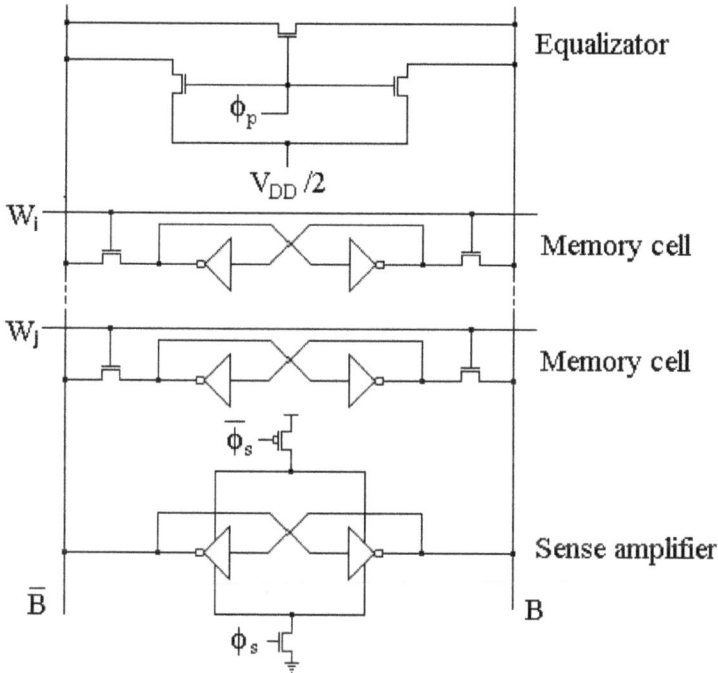

Fig. 5.24. The principal electrical circuit of a memory column.

obtain the defined voltages $V_B = V_{DD}$ and $V_{\bar{B}} = 0$. These voltages return the state of our virtual capacitors to the initial state and the information which was in the memory cell before reading process will be restored: $Q = $ "1" once more. So, the reading is a non-destructive process. Through the bit-lines this information will be transferred to the input-output terminal. The use of a sense amplifier significantly speeds up the process of reading information from a memory cell. A small difference of about $\Delta V = 120\,\text{mV}$ is enough to turn on the sense amplifier and get the information in the input/output terminal.

Figure 5.24 represents schematically one column from the memory net in more general form. The sense amplifier will also be switched on periodically. Such a connection enables us to decrease the power consumption. The column in Fig. 5.24 shows that one equalizer, one sense amplifier, and a lot of memory cells are connected in parallel within the column. The system works according to a control signal

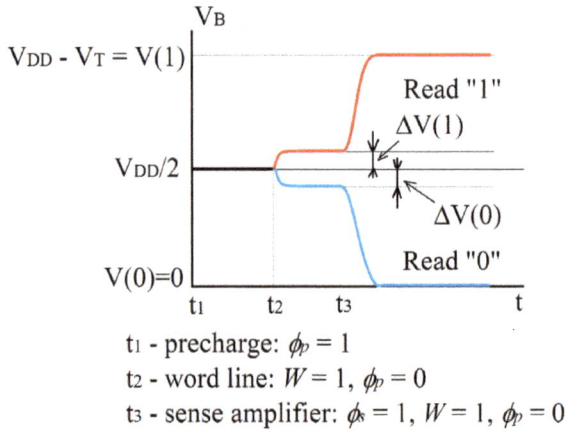

t_1 - precharge: $\phi_p = 1$
t_2 - word line: $W = 1$, $\phi_p = 0$
t_3 - sense amplifier: $\phi_s = 1$, $W = 1$, $\phi_p = 0$

Fig. 5.25.   A timing diagram of the reading process.

coming into the memory net. Figure 5.25 represents a timing diagram
of the voltage measured on the bit-line B.

The system works as follows:

1. The equalizator circuit charges the bit-lines up to the precharge
   voltage $V_{DD}/2$ after the short control signal $\phi_p = 1$.
2. The precharge signal is switched to zero. The row decoder chooses
   a suitable $W = 1$ and the reading process begins.
3. The sense amplifier is turned on with the control signal $\phi_s = 1$
   and simultaneously connects the bit-lines to the input/output
   terminal. In this order, the required information is read and
   refreshed in the memory cell.

The writing process occurs according to the same timing diagram.
However now the information comes from the input/output terminal
to the chosen bit-lines and charges the suitable memory cell according
to the chosen word-line through the chosen bit-line.

All processes in the memory cells take time. Let us analyze the
time required for reading the information from the memory cell.
Figure 5.26 illustrates our analysis.

Assume that this memory cell has the following data: $\mu_n C_{ox} =
50\,\mu A/V^2$, $\mu_p C_{ox} = 20\,\mu A/V^2$, $V_{DD} = 5\,V$, $V_{TN0} = -V_{TP0} = 1\,V$,

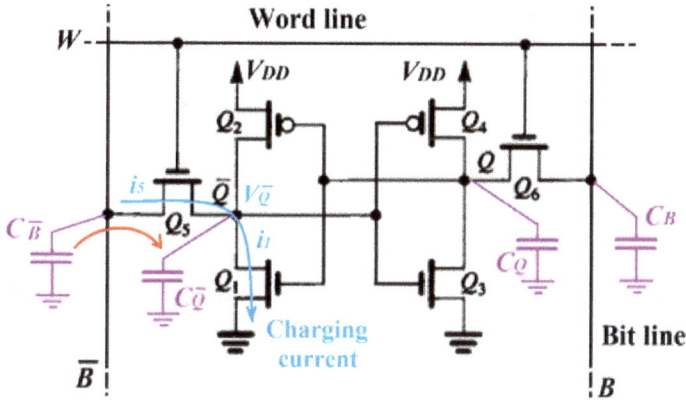

Fig. 5.26.   The currents flowing through the memory cell while reading.

$2\Phi_f = 0.6\,\text{V}$, $\gamma = 0.5\,\text{V}^{0.5}$, $\left(\frac{W}{L}\right)_n = \frac{4}{2}$, $\left(\frac{W}{L}\right)_{PTL,n} = \frac{10}{2}$, $\left(\frac{W}{L}\right)_p = \frac{10}{2}$, $C_B = 1\text{pF}$. Also assume for simplicity that this memory cell stores logic "1" or $V_Q = V_{DD}$ and both bit-lines are charged up to V_{DD}. Let us now calculate the time required to reach 0.2 V between bit-lines. In our calculation, it needs to be taken into account that the source and body of the transistor Q_5 are not connected and the voltage between source and body exists. This voltage acts on the threshold voltage of Q_5 as follows:

$$V_T = V_{T0} + \gamma(\sqrt{2\emptyset_f + V_{sb}} - \sqrt{2\emptyset_f}). \qquad (5.1)$$

Where $\gamma = \frac{\sqrt{2qN_A\varepsilon_s}}{C_{ox}}$ is a parameter of the fabrication process (see Eqs. (4.20) and (4.21)). According to our assumptions, after choosing the word-line W = "1", the right part of the circuit does not conduct a current such as $V_Q = V_B = V_{DD}$. At the same time, the bit-line $\bar{B}$ begins to charge the virtual capacitor $C_{\bar{Q}}$ through the two transistors Q_5 and Q_1 and a current i_5 flows in the right direction. The NMOS pass-transistor Q_5 with gate and drain connected to V_{DD} will be in the saturation mode and the transistor Q_1 with a small difference between its source and drain voltages will be in the triode mode of operation. Evidently, the current i_1 flowing through the transistor

Q_1 is equal to the i_5

$$i_1 = (\mu_n C_{ox}) \left(\frac{W}{L} \right)_n \left[(V_{DD} - V_{TN0})V_{\bar{Q}} - \frac{1}{2}V_{\bar{Q}}^2 \right]$$

$$= \frac{1}{2}(\mu_n C_{ox}) \left(\frac{W}{L} \right)_{PTL,n} (V_{DD} - V_{\bar{Q}} - V_T)^2 = i_5. \qquad (5.2)$$

This equation contains two variables: $V_{\bar{Q}}$ and V_T such as the source of Q_5 is separated from ground in the considered state. We can solve this equation using the iteration method. To do this, let us assume that $V_T = 1\,\text{V}$ in the first approximation. The substitution of this value into Eq. (5.2) in number form gives:

$$50 \cdot 10^{-6} \cdot \frac{4}{2} \left[(5-1)V_{\bar{Q}} - \frac{1}{2}V_{\bar{Q}}^2 \right] = \frac{1}{2} \cdot 50 \cdot 10^{-6} \cdot \frac{10}{2}(5 - V_{\bar{Q}} - 1).$$

$$(5.3)$$

The solution of this equation gives us the value $V_{\bar{Q}} = 1.86\,\text{V}$. Now, we will use this value in Eq. (5.1) which results in the second approximation $V_T = 1.4\,\text{V}$. The substitution of this value into Eq. (5.2) will give $V_{\bar{Q}} = 1.6\,\text{V}$. The third iteration does not provide an additional refinement in our case, therefore we can substitute the obtained values into Eq. (5.2) and calculate the current flowing through transistor Q_5. It equals to $i_5 \approx 0.5\,mA$ and this current is constant as it flows through the transistor in the saturation mode. The time required to decrease the voltage of the bit-line $V_{\bar{B}}$ from V_{DD} such that the $\Delta V = 0.2\,\text{V}$ is:

$$\Delta t = \frac{C_{\bar{B}} \cdot \Delta V}{i_5} = \frac{1 \cdot 10^{-12} \cdot 0.2}{0.5 \cdot 10^{-3}} = 0.4\,ns. \qquad (5.4)$$

5.3. Dynamic memory cells (DRAM)

5.3.1. *Four-transistor dynamic memory cell*

Figure 5.27 represents the principal electrical circuit of a dynamic memory organization based on the net concept. The main difference between this circuit and the static memory cell is in the application of the capacity of NMOS transistors. Here, a piece of information is

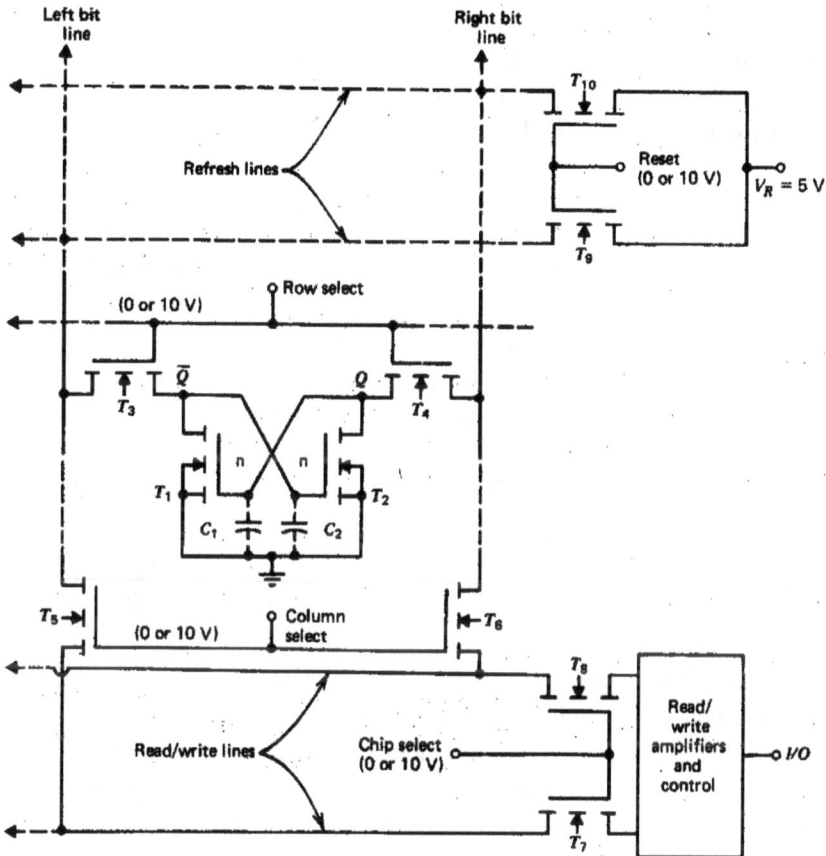

Fig. 5.27. The principal electrical circuit of a dynamic memory column.

stored in the virtual capacitors C_1 and C_2 which represents the input capacities of the transistors T_1 and T_2.

This circuit works according to the same time diagram as shown in Fig. 5.25. In the beginning of each operation period, both bit-lines are charged up to the voltage $V_{DD}/2 = 5\,\text{V}$ and the circuit waits for the choosing signal $W = 1$. The signal $W = 1$ initiates the reading process when the currents begin to flow in the memory cell and create the voltage difference between bit-lines. The signal of the chosen column connects the cell with the sensitive amplifier and transfers bit-lines with the input/output terminal. With the help of the sensitive amplifier, a piece of information is restored in the cell. The main requirement of this system is as follows: the period taken to complete the read/write operation should be shorter than the time taken for information to be lost through the leakage currents.

5.3.2. *One-transistor dynamic memory cell*

The most commonly used modern dynamic memory is based on the net-shaped one transistor scheme. The information bit stored in the transistor is utilized as a capacitor between the gate and the source. This capacitor is connected with the bit-line through a pass transistor of the N-type. A required memory cell may be chosen using a suitable word-line and applied to the bit-line. A sense amplifier is also connected to the bit-line. This scheme is presented in Fig. 5.28(a).

As shown, a pass-transistor T connects a memory cell (the capacitor C_M) with the word-line and with the bit-line, making it possible to write an information bit in this capacitor and to read it. A capacitor C_B represents a bit-line capacity. This capacitor is coloured blue since it is virtual and since it represents the capacitance of a bit-line. Usually, the C_B capacity is more than the C_M capacity by 30–50 times. The read-out process here occurs in the same way as in the case of the static RAM. In other words, we have a three-step process with the same control signals: ϕ_p, W, ϕ_s. The system also works as follows:

1. The equalizator circuit charges the bit-lines up to the precharge voltage $V_{DD}/2$ after the short control signal $\phi_p = 1$.

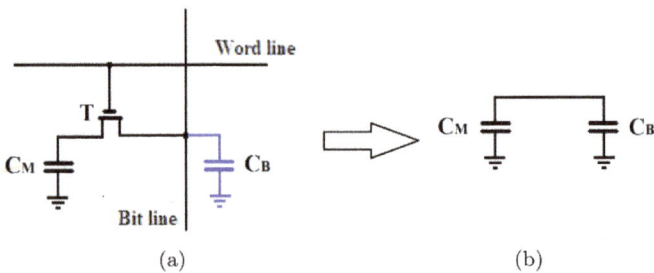

Fig. 5.28. The principal electrical circuit of a one-transistor dynamic memory net.

2. The precharge signal is switched to zero. The row decoder chooses a suitable $W = 1$ and the reading process begins.
3. The sense amplifier is turned on with the control signal $\phi_s = 1$ and simultaneously connects the bit-lines to the input/output terminal. In this order, the required information is read and refreshed in the memory cell.

However, in this case, there are several differences due to the one bit-line organization. Let us consider the reading process. The first control signal, $\phi_p = 1$, charges the bit-line up to a voltage equal to $V_{DD}/2$. When the control signal for row selection will be chose the $W = 1$, the circuit shown in Fig. 5.28(a) will be converted as shown in Fig. 5.28(b). Now, the reading process begins. During this process, a common charge will be obtained in two capacitors, C_M and C_B, connected in parallel. If we assume that the voltage V_M is stored in the capacitor C_M representing the memory cell, according to the charge storage law, the reading process may be written as follows:

$$C_M V_M + C_B \frac{V_{DD}}{2} = (C_B + C_M) \left(\frac{V_{DD}}{2} + \Delta V \right). \quad (5.5)$$

From this equation we obtain

$$\Delta V = \frac{C_M}{C_B + C_M} \left(V_M - \frac{V_{DD}}{2} \right) \quad \text{and such as} \quad C_B \gg C_M,$$

$$\Delta V = \frac{C_M}{C_B} \left(V_M - \frac{V_{DD}}{2} \right). \quad (5.6)$$

We have two possibilities for the voltage V_M: (1) $V_M = $ "1" $= V_{DD} - V_{TN}$ and (2) $V_M = $ "0" $= 0$. These possibilities relate with logical "1" and "0" stored in the memory cell. Therefore, according to the content of the memory cell, Eq. (5.6) will transform as follows:

1. $V_M = $ "1" $= V_{DD} - V_{TN}$: $\quad \Delta V_1 = \frac{C_M}{C_B} \left(\frac{V_{DD}}{2} - V_{TN} \right). \quad (5.7)$

2. $V_M = $ "0" $= 0$: $\quad \Delta V_0 = -\frac{C_M}{C_B} \cdot \frac{V_{DD}}{2}. \quad (5.8)$

Due to the small value of the C_M/C_B relation, the read-out voltage will be low. Therefore, the reading rate will be more than in the case of the static RAM. For example, if we take $V_{DD} = 5$V, $V_{TN} = 1$V and $C_M/C_B = 0.02$, the voltage difference will be equal to $\Delta V_0 = -50$ mV and $\Delta V_1 = 30$ mV. This voltage will control the sense amplifier to read the right information bit from the memory cell and to renew this information in the memory cell.

5.4.   Memory related circuitry

5.4.1.   *The row-address decoder*

Let us consider once more the net structure of the random-access memory presented in Fig. 5.29. As shown in this figure, there are

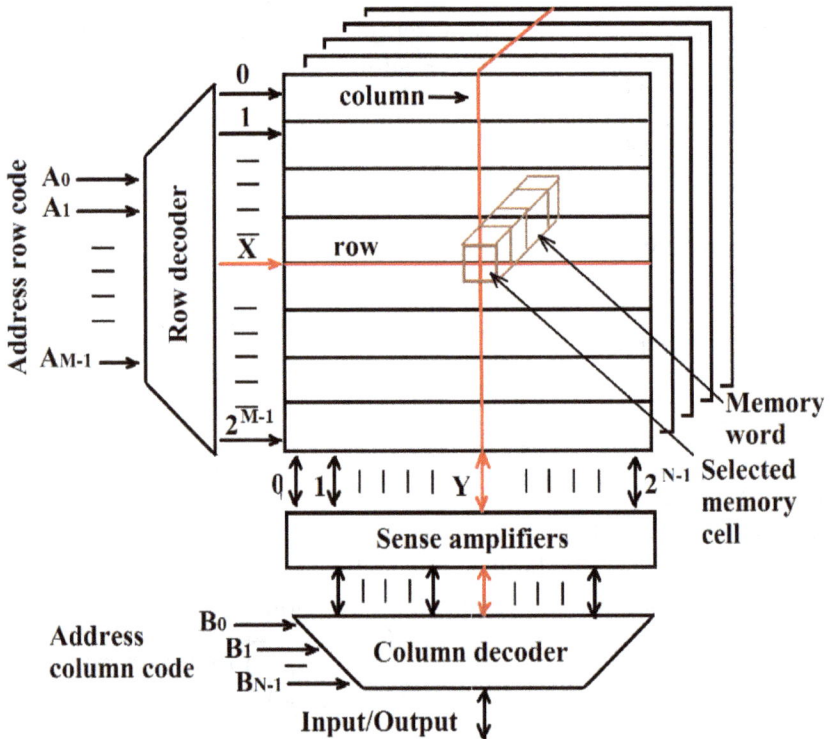

Fig. 5.29.   The principal scheme of the memory net organization.

additional circuits which make it possible to apply this model to each memory cell with the same rate. These circuits are the row decoder and the column decoder.

To understand how these circuits work, we should consider the inside construction of these schemes. The main requirement of these circuits is that the circuits should be combinational and simple.

According to the definition, a decoder is a circuit which has a number of inputs k and a number of outputs n connected in such a way that $n = 2^k$ and each output is defined using only one combination of input signals, when both the input and output signal can take only one value out of two: logic 1 and logic 0. Figure 5.30 represents a truth table built according to this definition. From this truth table, one can write Boolean equations for each row:

$$W_0 = \overline{A_0} \cdot \overline{A_1} \cdot \overline{A_2} = \overline{A_0 + A_1 + A_2}.$$

$$W_1 = A_0 \cdot \overline{A_1} \cdot \overline{A_2} = \overline{\overline{A_0} + A_1 + A_2}.$$

$$W_0 = \overline{A_0} \cdot A_1 \cdot \overline{A_2} = \overline{A_0 + \overline{A_1} + A_2}.$$

All these equations describe a circuit of the NOR type.

So, we obtain the consequence of NOR circuits arranged together, as shown in Fig. 5.31.

This circuit works as follows: each input variable can take one of two possible values, "0" or "1". When the control signal $\phi_p = 0$,

	A2	A1	A0	W0	W1	W2	W3	W4	W5	W6	W7
W0	0	0	0	1	0	0	0	0	0	0	0
W1	0	0	1	0	1	0	0	0	0	0	0
W2	0	1	0	0	0	1	0	0	0	0	0
W3	0	1	1	0	0	0	1	0	0	0	0
W4	1	0	0	0	0	0	0	1	0	0	0
W5	1	0	1	0	0	0	0	0	1	0	0
W6	1	1	0	0	0	0	0	0	0	1	0
W7	1	1	1	0	0	0	0	0	0	0	1

Fig. 5.30. The true-table of the row decoder.

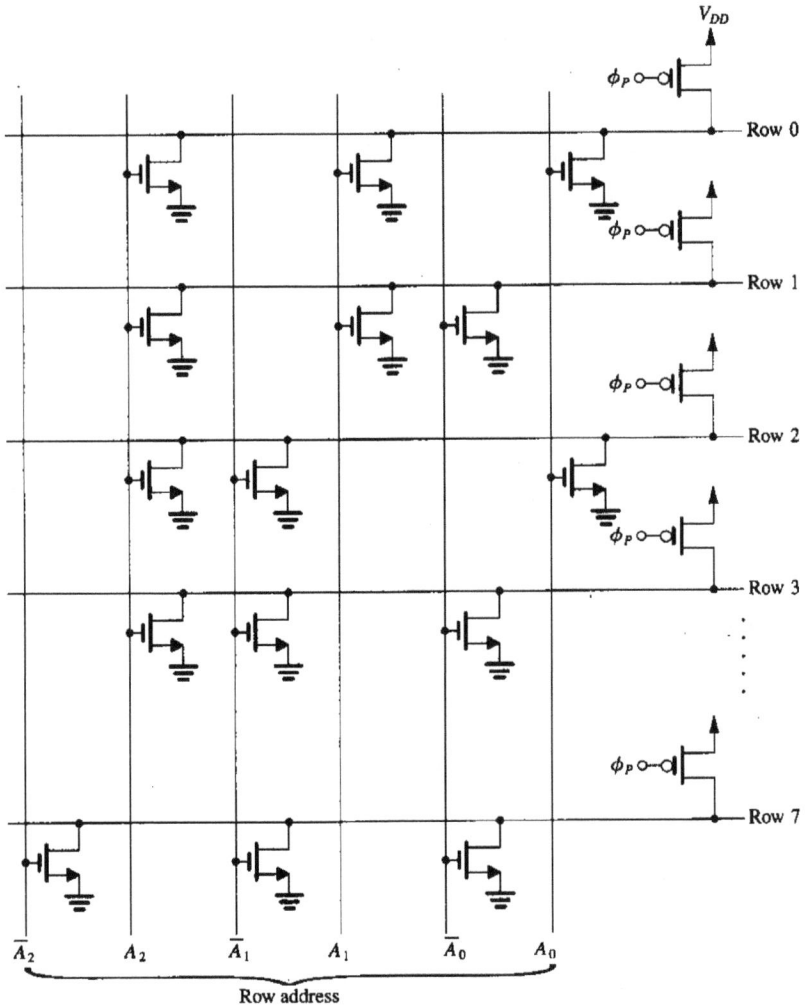

Fig. 5.31.   The sequence of NOR circuits arranged together and playing the role of the NOR decoder.

through the precharge time, all the rows are charged up to V_{DD}. When the control signal changes the state on $\phi_p =$ "1", the evaluation time, only one row, according to the suitable code, will store the logic state "1" and all other rows will be discharged. Thus, according to the inserted code, the row will be chosen.

5.4.2. The column decoder

The main column decoder used is also based on the NOR decoder scheme. However, this NOR decoder is integrated in one module with the pas-transistor multiplexer so that it can be applied to columns. The principal electric circuit of a column decoder, consisting of a NOR decoder and a pas-transistor, is presented in Fig. 5.32.

As can be seen from Fig. 5.32, the column decoder integrates two technologies in the same device: a NOR decoder and a PTL multiplexer. Each chosen row on the output of the ROW decoder promotes a high voltage (logic "1") on a suitable gate of the PTL transistor, getting on the required column and connecting it with the input/output terminal. In this way, a code on the input of the ROW decoder defines the suitable column chosen.

Another type of column decoder is presented in Fig. 5.33. This circuit is called the three-decoder. In this circuit, a code for the required column is inserted into a tree built from 2:1 multiplexers. By this method, we can decrease the number of used transistors, however, in this scheme, the signal propagation time increases significantly.

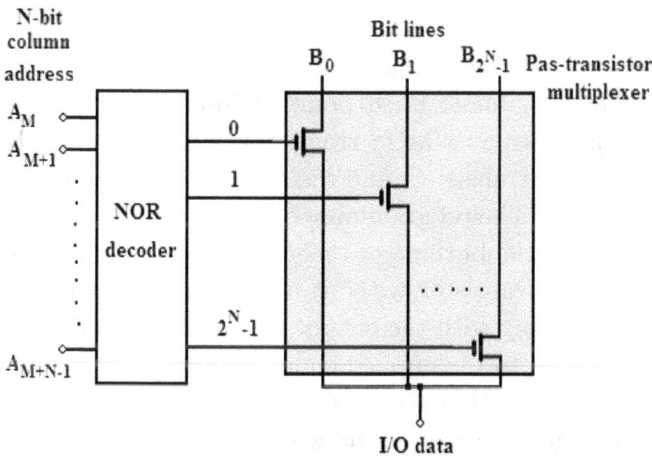

Fig. 5.32. The principal electric circuit of the column decoder.

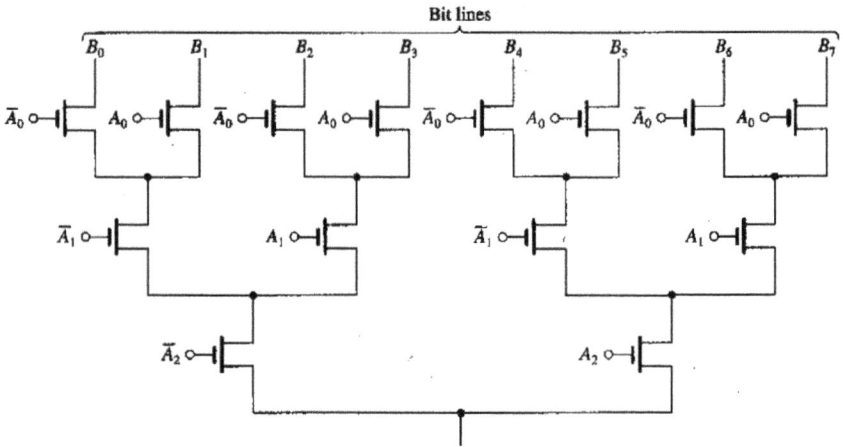

Fig. 5.33.   The principal electric circuit of a column decoder of the three-type.

5.5.   Read-only memory (ROM)

Sometimes, information should be stored without allowing us to change it. To achieve this goal, the read-only memory (ROM) was developed. ROM was created to store significant information which should be stored without any changes throughout the lifetime of the device. A simple scheme explaining read only memory is shown in Fig. 5.34. This figure represents the circuit which consists of the row decoder and the net with the stored information. This net is built from rows and columns. Each point of intersection of a row with a column concludes an NMOS transistor. Rows are connected with the gates of these transistors and their drains are connected with the suitable columns. The programming of this memory or the writing of information is in the burning of the suitable gates of the transistors. In this case, the transistor without a gate cannot connect the logic "1" from the supply with the output terminal. This circuit works as follows: according to the applied code, a suitable row will be chosen and all transistors with gates connected to this row will show "0" on the terminals, however transistors which have burned gates will show "1". So, the terminal row will present the row stored in the memory according to the chosen code.

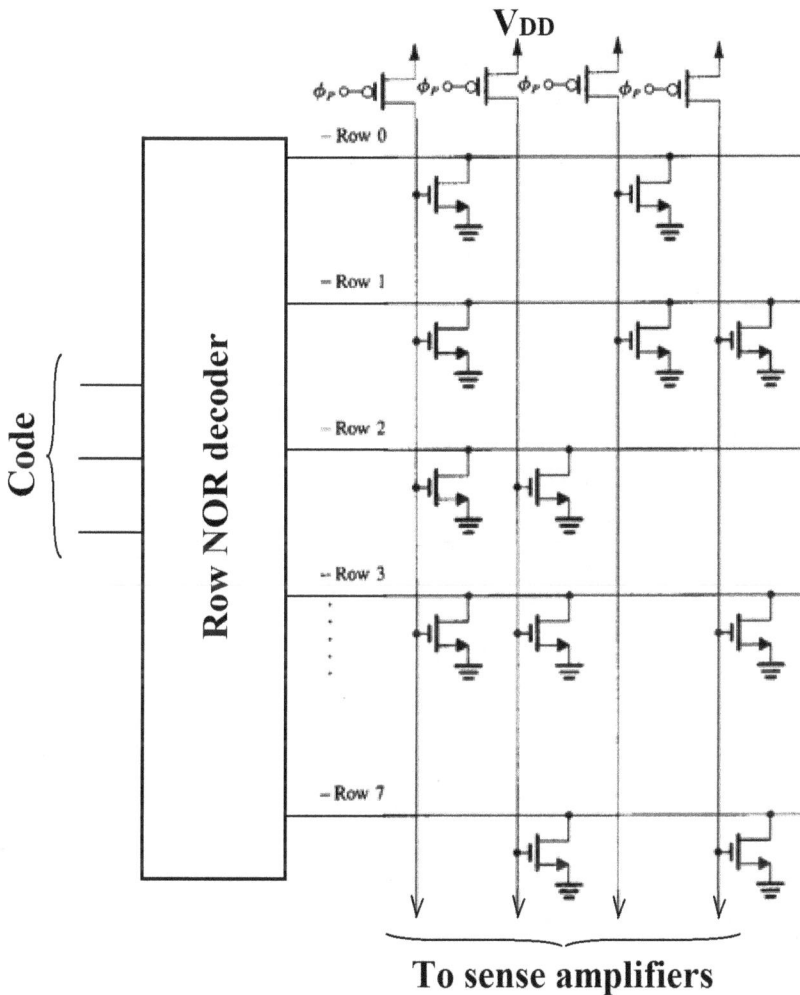

Fig. 5.34. The read-only memory circuit.

5.6. Problems

5.6.1 It is known that the capacity of a memory cell in the DRAM system is equal to 0.1 fF, the supply voltage is 1.2 V and the refresh information period is 1 ms. Calculate the leakage current which fully discharges the memory cell during the defined period.

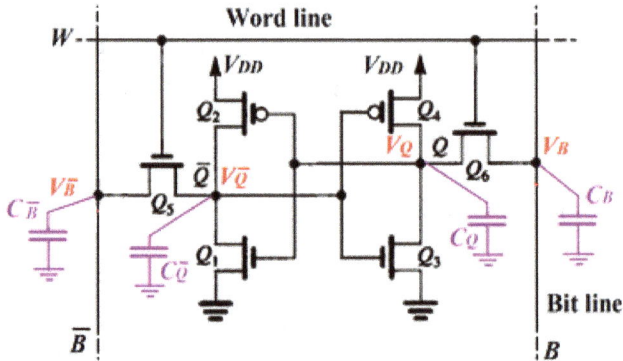

Fig. 5.35. A SRAM memory cell.

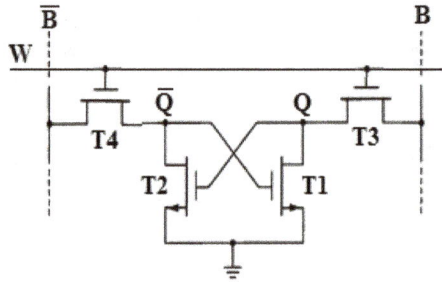

Fig. 5.36. A NMOS-DRAM memory cell.

5.6.2 Program a MOS ROM memory consisting of 8 rows and 4 columns: (msb)**A802BF45**.

5.6.3 Figure 5.35 represents a SRAM memory cell.

For this cell, the following data is known: $V_{TP} = -1\,\mathrm{V}$, $V_{TN} = 1\,\mathrm{V}$, $K'_p = 20\,\mu\mathrm{A/V2}$, $K'_n = 40\,\mu\mathrm{A/V2}$, $(W/L)_{Q1} = (W/L)_{Q3} = 2$, $(W/L)_{Q2} = (W/L)_{Q4} = 4$, $(W/L) = 1$ for all other transistors, $V_{DD} = 5\,\mathrm{V}$, and the bit-lines are charged up to V_{DD}.

If $Q = 0$ and $\bar{Q} = 1$, calculate the voltage at the point Q after applying the voltage V_{DD} to the word-line.

5.6.4 Figure 5.36 represents a memory cell of the NMOS-DRAM type.

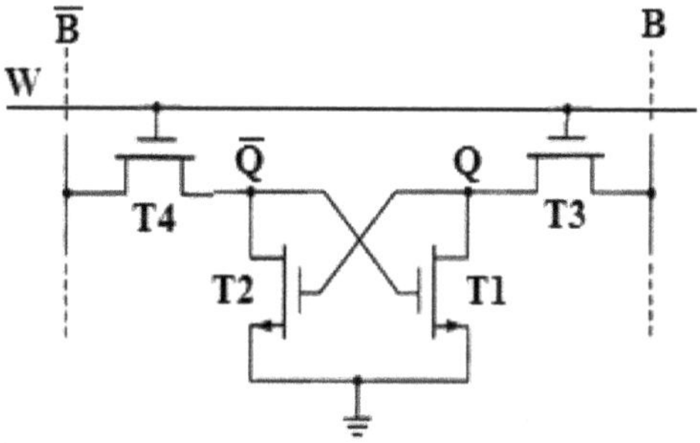

Fig. 5.37.   A NMOS-DRAM memory cell.

It is known that: $\gamma = 0.5V^{0.5}$, $2\Phi_f = 0.6\,V$, $(W/L)_N = 2$, $V_{TN} = 0.8\,V$, $k'_n = 40\,\mu A/V^2$, $V_{dd} = 4\,V$, the length of each transistor is the same, $L = 0.6\,\mu m$, Q = "1", $\bar{Q}$ = "0", and electron mobility is $\mu_n = 1000\,cm^2/V{\cdot}s$. If $B = \bar{B} = V_{DD}/2$ and W = "1", calculate how much time is needed to obtain $\Delta V = V_B - V_{\bar{B}} = 50\,mV$.

5.6.5 Figure 5.37 represents a memory cell of the NMOS-DRAM type.

It is known that: $(W/L)_N = 5/2$, $V_{TN} = 0.4\,V$, $k'_n = 20\,\mu A/V^2$, $V_{dd} = 2\,V$, the length of each transistor is the same, $L = 0.6\,\mu m$, Q = "0", $\bar{Q}$ = "1", and electron mobility is $\mu_n = 1000\,cm^2/V{\cdot}s$.

Calculate how much time is needed to obtain Q = "1" from the moment when W = "1".

Chapter 6

Solutions

2.8.1 The diode behavior is defined by the Shockley model (see Eq. (2.1)):

$$I = I_s \left(e^{\frac{V_a}{V_t}} - 1 \right).$$

Differentiating this equation will give the definition of the internal resistance:

$$dI = \frac{I_s}{V_t} e^{\frac{V_a}{V_t}} dV_a; \quad r_\gamma = \frac{dV_a}{dI} = \frac{V_t}{I_s} e^{-\frac{V_a}{V_t}}.$$

At temperature $T = RT$, $V_t = 0.026\,\text{V}$, $I_s = 10^{-14}\,\text{A}$.

Calculations are presented in Table 6.1:

Table 6.1. Solution of the exercise.

V_a, V	0.4	0.5	0.6	0.7
r_γ, Ω	541401	11565	247	5.3

2.8.2 The capacity of the PN junction is defined by the following equation (Eq. (2.5)):

$$C_j = \frac{\varepsilon_0 \varepsilon_r A}{W} = A \sqrt{\frac{q \varepsilon_0 \varepsilon_r N_A N_D}{2(N_A + N_D)(V_b - V_a)}}.$$

The built-in potential of the junction at room temperature is (see Eq. (2.6)):

$$V_b = V_t ln \frac{N_D N_A}{n_i^2} = 0.026 ln \frac{4 \cdot 10^{17} 4 \cdot 10^{15} \, 20}{2.25 \cdot 10} = 0.77 \, V.$$

Without external voltage, the capacity will be equal to:

$$C_j = 10^{-6} \sqrt{\frac{1.6 \cdot 10^{-19} 8.85 \cdot 10^{-14} 11.7 \cdot 4 \cdot 10^{17} 4 \cdot 10^{15}}{2 \cdot 10^{17} 0.77}} = 41.5 \, fF.$$

2.8.3 According to the definition (see Eq. (2.4)) and using the Shockley model, we can calculate the diffusion capacity under the external direct voltage of 0.6 V:

$$C_D = \frac{dQ}{dV_a} = \frac{d \left(i_D t \right)}{dV_a} = \tau \frac{dI_s \left(e^{\frac{V_a}{V_t}} - 1 \right)}{dV_a} = \frac{\tau I_s}{V_t} e^{\frac{V_a}{V_t}}.$$

Here, the leakage current I_s (see Eq. (2.2)) may be calculated as follows:

$$I_s = q A n_i^2 \left(\frac{D_n}{L_n N_D} + \frac{D_p}{L_p N_A} \right).$$

The diffusion coefficient we find using the Einstein's relation

$$D_n = \mu_n V_t = 400 \cdot 0.026 = 10.4 \frac{cm^2}{V \cdot s}.$$

The diffusion length (see Eq. (2.3)) $L_n = \sqrt{D_n \tau_n} = \sqrt{10.4 \cdot 0.1} = 1.02 \, cm$.

$$I_s = 1.6 \cdot 10^{-19} 10^{-6} 2.25 \cdot 10^{20} \frac{10.4}{1.02 \cdot 10^{15}} = 4.9 \cdot 10^{-19} \, A.$$

$$C_D = \frac{\tau I_s}{V_t} e^{\frac{V_a}{V_t}} = \frac{0.1 \cdot 4.9 \cdot 10^{-19}}{0.026} e^{\frac{0.6}{0.026}} \approx 20 \, nF.$$

2.8.4 The presented digital circuit can be in two different states, $V_{out} = $ "0" and $V_{out} = $ "1".

If the input voltage $V_{in} = $ "0", the transistor is in the cutoff state, the currents in both circuits are zero and the dissipated power is zero ($P_0 = 0$). If the input voltage is "1", the transistor is in the saturation state and we have currents in the base circuit and in the collector circuit. The dissipated power will be the sum of the power dissipated in both resistors, Rb and Rc.

Therefore, $P_1 = V_{cc}\left(\frac{V_{cc}}{R_b} + \frac{V_{cc}}{R_c}\right) = 5\left(\frac{5}{10000} + \frac{5}{1000}\right) = 27.5\,mW$.

Average power: $Pav = 0.5(0 + 27.5) = 13.75\,mW$.

2.8.5 We will calculate the required parameters according to Eqs. (2.5) and (2.6) and the charge conservation law for semiconductors ($n_i^2 = N_A N_D$) at room temperature.

a. According to the data presented in the exercise,

$n_{nE} = N_{DE} = 5 \cdot 10^{17}\,cm^{-3}$, $p_{nE} = n_i^2/n_{nE} = 2.25 \cdot 10^{20}/5 \cdot 10^{17} = 4.5 \cdot 10^3\,cm^{-3}$;

$p_{pB} = 10^{16}\,cm^{-3}$, $n_{pB} = n_i^2/p_{pB} = 2.25 \cdot 10^{20}/10^{16} = 2.25 \cdot 10^4\,cm^{-3}$;

$n_{nC} = 10^{15}\,cm^{-3}$, $p_{nC} = n_i^2/n_{nC} = 2.25 \cdot 10^{20}/10^{15} = 2.25 \cdot 10^5\,cm^{-3}$.

b. Built-in potential:

$$V_{BE} = V_t ln \frac{N_D N_A}{n_i^2} = 0.026 ln \frac{5 \cdot 10^{17} 10^{16}}{2.25 \cdot 10^{20}} = 0.8\,V$$

$$V_{BC} = V_t ln \frac{N_D N_A}{n_i^2} = 0.026 ln \frac{10^{15} 10^{16}}{2.25 \cdot 10^{20}} = 0.64\,V.$$

Depletion zones of the base-emitter junction without any external voltage:

$$x_{B0} = \sqrt{\frac{2\varepsilon_0\varepsilon_r V_{BE} N_A}{q N_D(N_A + N_D)}}$$

$$= \sqrt{\frac{2 \cdot 8.85 \cdot 10^{-14} 11.7 \cdot 0.8 \cdot 5 \cdot 10^{17}}{1.6 \cdot 10^{-19} \cdot 10^{16}(5 \cdot 10^{17} + 10^{16})}} = 320\,nm.$$

$$x_{E0} = \sqrt{\frac{2\varepsilon_0\varepsilon_r V_{BE} N_D}{q N_A (N_A + N_D)}}$$

$$= \sqrt{\frac{2 \cdot 8.85 \cdot 10^{-14} 11.7 \cdot 0.8 \cdot 5 \cdot 10^{16}}{1.6 \cdot 10^{-19} 5 \cdot 10^{17} (5 \cdot 10^{17} + 10^{16})}} = 14\, nm.$$

Depletion zones of the base-collector junction without any external voltage:

$$x_{B0} = \sqrt{\frac{2\varepsilon_0\varepsilon_r V_{BE} N_A}{q N_D (N_A + N_D)}}$$

$$= \sqrt{\frac{2 \cdot 8.85 \cdot 10^{-14} 11.7 \cdot 0.64 \cdot 10^{16}}{1.6 \cdot 10^{-19} \cdot 10^{15} (5 \cdot 10^{17} + 10^{16})}} = 127\, nm.$$

$$x_{C0} = \sqrt{\frac{2\varepsilon_0\varepsilon_r V_{BE} N_D}{q N_A (N_A + N_D)}}$$

$$= \sqrt{\frac{2 \cdot 8.85 \cdot 10^{-14} 11.7 \cdot 0.64 \cdot 10^{15}}{1.6 \cdot 10^{-19} \cdot 10^{16} (5 \cdot 10^{17} + 10^{16})}} = 13\, nm.$$

c. The following Fig. 6.1 represents the exercise's data:

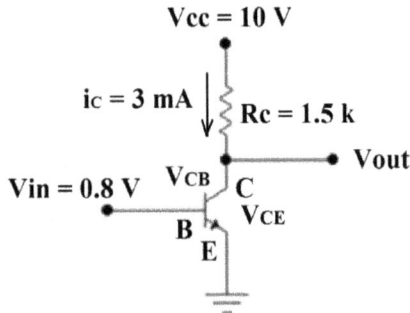

Fig. 6.1.   The exercise's data.

Voltage $V_{CE} = Vcc - i_C Rc = 10 - 3 \cdot 1.5 = 5.5\, V.$
Voltage $V_{CB} = V_{CE} - Vin = 5.5 - 0.8 = 4.7\, V.$

2.8.6 The right and simple way to identify the type of logic gate is to build the truth table (see Table 6.2).

Table 6.2.   Truth table.

A	B	F
0	0	1
0	1	0
1	0	0
1	1	0

a. This is the **NOR** function.
b. If both inputs are low, we have $V_{out} =$ "1" $= Vcc = 5\,V$.

2.8.7 To calculate the noise margins, we need to build the transfer function for the logic gate. First of all, we short-circuit all the inputs together to obtain the inverter.

1. Now, we bring the input voltage to all inputs and we begin from $V_{in} =$ "0". In this case, all the transistors are in the cutoff state and $V_{out} =$ "1" $= Vcc$.
2. If the input voltage is high, $V_{in} =$ "1" $= V_{cc}$, the transistor T1 will be in the active mode and the transistors T2 and T3 will be in the saturation mode.
3. Changes in the output voltage begin during the transition of the transistor T3 from the cutoff state to the active state, that requires $V_{BE3} = V_{BE,\gamma} = 0.5\,V$. If we take into account that the transistor T1 is in the active state, we can find that

$$Vin1 = V_{IL} = V_{BE3,\gamma} + V_{BE2,on} + V_{BE1,\gamma} - V_{D,on} - V_{"0"}$$

or

$V_{IL} = 0.6 + 0.7 + 0.5 - 0.7 - 0.2 = 0.9\,V$.

Transistor T3 enters the saturation mode when VBE3 $= 0.7$ V. Therefore, $V_{IH} = V_{BE3,on} + V_{BE2,on} + V_{BE1,\gamma} - V_{D,on} - V_{"0"} = 0.7 + 0.7 + 0.6 - 0.7 - 0.2 = 1.1\,V$.

$V_{OH} = V_{cc} = 5\,V$, $V_{OL} = V_{CE,sat} = 0.2\,V$, therefore:

$N_{ML} = V_{IL} - V_{OL} = 0.9 - 0.2 = 0.7\,V$.
$N_{MH} = V_{OH} - V_{IH} = 5 - 1.1 = 3.9\,V$.

2.8.8 To identify the logic function, we build a truth table (Table 6.3):

Table 6.3.   Truth table.

N	A	B	C	D	Vo
1	0	0	0	0	1
2	0	0	0	1	1
3	0	0	1	0	1
4	0	0	1	1	0
5	0	1	0	0	1
6	0	1	0	1	1
7	0	1	1	0	1
8	0	1	1	1	0
9	1	0	0	0	1
10	1	0	0	1	1
11	1	0	1	0	1
12	1	0	1	1	0
13	1	1	0	0	0
14	1	1	0	1	0
15	1	1	1	0	0
16	1	1	1	1	0

a. To find the logic function we build the Karnaugh map (Table 6.4):

Table 6.4.   The Karnaugh map.

AB\CD	00	01	11	10
00	1	1	0	1
01	1	1	0	1
11	0	0	0	0
10	1	1	0	1

$$F = A'C' + A'D' + B'C' + B'D' = (A'+B')(C'+D').$$

b. At state 1 in the truth table, only the input currents through resistors R1 and R3 flow in Fig. 6.2.

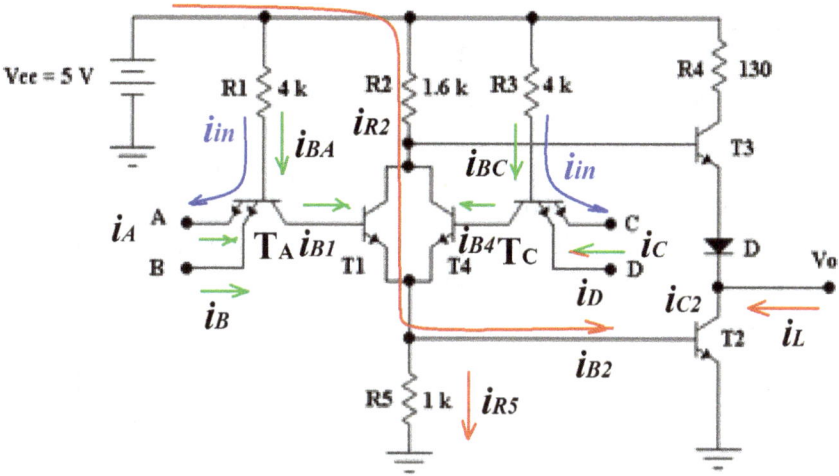

Fig. 6.2. The behavior of currents in the circuit.

These currents may be calculated using Kirchhoff's equations. When all inputs are in the low state:

$$i_{in} = 0.5\frac{V_{cc} - V_{BEA,on} - V_{\text{``0''}}}{R1} = 0.5\frac{5 - 0.7 - 0.2}{4} = 0.51\,mA.$$

The power in this state is $P1 = 2 \cdot i_{BA} \cdot V_{cc} = 2 \cdot 1.025 \cdot 10 = 10.25\,\text{mW}$.

In the case when all inputs are in the high state, we have currents in resistors R1, R2, R3, and R5.

$$i_{R5} = \frac{V_{BE2,on}}{R5} = \frac{0.7}{1} = 0.7\,mA$$

$$i_{BA} = i_{BC} = \frac{V_{cc} - V_{BEA,on} - V_{BE1,on} - V_{BE2,on}}{R1}$$

$$= \frac{5 - 2.1}{4} = 0.725\,mA$$

$$i_{R2} = \frac{V_{cc} - V_{CE1,sat} - V_{BE2,on}}{R2} = \frac{5 - 0.2 - 0.7}{1.6} = 2.56\,mA.$$

The full current $i_\Sigma = 0.7 + 2 \cdot 0.725 + 2.56 = 4.71\,mA$, P0= $i_\Sigma \cdot V_{cc} = 4.71 \cdot 5 = 23.55\,mW$.

Pav $= 0.5(10.25 + 23.55) = 16.9$ mW.

c. The maximum fan-out we calculate is in the case Vo = "0" when $A = B = C = D = V_{\text{"0"}}$. Now, both transistors T_A and T_B are in the reverse active mode.

The point at which the transistor T2 transitions from the saturation mode to the active mode will affect the relation of magnification for the transistor. Therefore, this will be as follows:

$$\beta_F i_{B2} = i_{C2} = i_L = N i_{in}, \text{ therefore } N_{max} = Int \left\{ \frac{\beta_F i_{b2}}{i_{in}} \right\}$$

$$i_{B2} = i_{R2} + i_{B1} + i_{B4} - i_{R5}$$

$$i_{B1} = i_{B4} = i_{BA} + i_A + i_B = i_{BA}(1 + 2\beta_R)$$

$$= 0.725 \cdot (1 + 0.2) = 0.87\,mA$$

$$i_{B2} = 2.56 + 2 \cdot 0.87 - 0.7 = 3.6\,mA$$

$$N_{max} = Int \left\{ \frac{\beta_F i_{b2}}{i_{in}} \right\} = Int \left\{ \frac{50 \cdot 3.6}{0.51} \right\} = Int\{352.94\} = 352.$$

2.8.9 To identify the logic function, we build the truth table shown in Fig. 6.3.

a. Input C is the "enable" input. If this input is low (C = "0"), the circuit is in the "third" state with the high impedance. A and B are logic inputs.

b. We calculate the input current for the state A = 1, B = 0, C = 1. In this case, the input current flows through the input B:

$$i_{in} = \frac{V_{cc} - V_{BE1,on} - V_{\text{"0"}}}{R1} = \frac{5 - 0.7 - 0.2}{2.8} = 1.46\,mA.$$

We calculate the maximum fan-out for the state A = B = C = 1. In this state, transistors T1 and T6 are in the reverse active state,

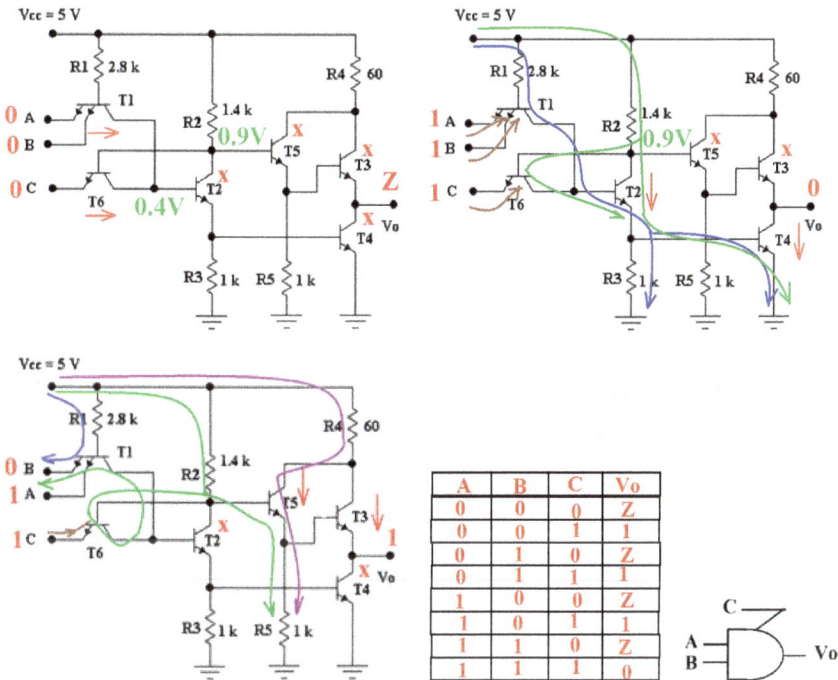

Fig. 6.3. The behavior of the logic circuit and the truth table of the implemented function.

transistors T2 and T4 are in the saturation state, and transistors T5 and T3 are in the cutoff state.

$$i_{B4} = i_{E2} - i_{R3}, \quad i_{E2} = i_{B2} + i_{C2},$$

$$i_{R3} = \frac{V_{BE4,on}}{R3} = \frac{0.7}{1} = 0.7\,mA$$

$$i_{C2} = \frac{V_{cc} - V_{CE2,sat} - V_{BE4,on}}{R2} = \frac{5 - 0.2 - 0.7}{1.4} = 2.93\,mA$$

$$i_{B2} = i_{C1} + i_{C6}, \quad i_{C1} = i_{B1}(1 + 2\beta_R), \quad i_{C6} = i_{B6}(1 + \beta_R)$$

$$i_{B1} = \frac{V_{cc} - V_{BE1,on} - V_{BR2,on} - V_{BE4,on}}{R1} = \frac{5 - 2.1}{2.8} = 1.01\,mA$$

$$i_{C1} = 1.01(1 + 0.2) = 1.24\,mA, \quad i_{B6} = \frac{5 - 2.1}{1.4} = 2.1\,mA$$

$$i_{C6} = 2.1 \cdot 1.1 = 2.28\,mA, \quad i_{B2} = 1.24 + 2.28 = 3.52\,mA,$$

$$i_{E2} = 2.93 + 3.52 = 6.45\,mA, \quad i_{B4} = 6.45 - 0.7 = 5.75\,mA$$

$$\beta_F i_{B4} = i_{C4} = i_L = N i_{in},$$

$$N_{max} = Int\left\{\frac{\beta_F i_{b4}}{i_{in}}\right\} = Int\left\{\frac{50 \cdot 5.75}{1.46}\right\} = Int\{196.91\} = 196.$$

2.8.10 To identify the logic function, we build a truth table (Table 6.5).

Table 6.5.   Truth table.

Vx	Vy	V1	V2
0	0	1	0
0	1	0	1
1	0	0	1
1	1	0	1

a. The circuit implements the functions **OR/NOR**.
b. Assume that Vx = Vy = "1", that the current $I_E = I_{C1}$ and that this current flows through transistors Q1 and Q2. Now, one can find a voltage V_E: $V_E = V_x - V_{BE1,on} = 4.5 - 0.7 = 3.8\,V$. The current $I_{C1} = I_E = V_E/R2 = 3.8/1.18 = 3.22\,mA$.

Assume that Vx = Vy = "0", that the current $I_E = I_{C2}$ and that this current flows through transistor Q3.

In this case, the voltage $V_E = V_3 - V_{BE3,on} = 4.15 - 0.7 = 3.45\,V$. The current $I_{C2} = I_E = V_E/R2 = 3.45/1.18 = 2.92\,mA$.

To obtain the symmetric behavior of the circuit, it is necessary for $I_{C1}R_{C1} = I_{C2}R_{C2} = V_{BE,on}$, therefore:

$$R_{C1} = \frac{0.7 \cdot 10^3}{2.92} = 240\,\Omega, \quad R_{C1} = \frac{0.7 \cdot 10^3}{3.22} = 217\,\Omega.$$

c. From the truth table, at the state Vx = Vy = "0", V1 = "1" and V2 = "0". V2 = "0" = $V_{cc} - I_{C2} - V_{BE5,on} = 5.2 - 0.7 - 0.7 = 3.8\,V$.

3.5.1 To identify the logic function, we build a truth table (Table 6.6).

Table 6.6. Truth table.

A	B	V_{out}
0	0	1
0	1	1
1	0	1
1	1	0

a. The circuit implements the function **NAND**.
b. (i) A = "0", B = "1". In this case, transistor $Q2$ is in the cutoff state. Now, $V_{out,1} = Vdd = 5\,V$.
(ii) A = B = "1". In this case, transistor Q1 is in the saturation mode and both transistors Q2 and Q3 are in the active state.

$$K_D[2(V_{DD} - V_{TND})V_{outL} - V_{outL}^2] = K_L V_{TNQ1}^2$$

$$= \mathrm{Idl} = K_n' \left(\frac{W}{L}\right)_{Q1} V_{TNQ1}^2$$

$$\mathrm{Idl} = \mu_n C_{ox} \left(\frac{W}{L}\right)_{Q3} V_{TNL}^2 = 35 \cdot 10^{-6}(1-2)^2 = 0.14\,\mathrm{mA}.$$

The resistance of transistors $Q2$ and $Q3$ in the conductive state is:

$$R_{onQ2} = \frac{1}{K_n(V_{DD} - V_T)} = \frac{1}{(\mu_n C_{ox})\left(\frac{W}{L}\right)_{Q2}(V_{DD} - V_{TNQ2})}$$

$$= \frac{1}{35 \cdot 10^{-6}4 \cdot 4.2} = 1.7\,k\Omega$$

$$R_{onQ3} = \frac{1}{K_n(V_{DD} - V_T)} = \frac{1}{(\mu_n C_{ox})\left(\frac{W}{L}\right)_{Q3}(V_{DD} - V_{TNQ3})}$$

$$= \frac{1}{35 \cdot 10^{-6}4 \cdot 4.2} = 1.7\,k\Omega$$

$$R_\Sigma = R_{onQ2} + R_{onQ3} = 1.7 + 1.7 = 3.4\,k\Omega,$$

$$V_{out,0} = \mathrm{Idl} \cdot R_\Sigma = 0.14 \cdot 3.4 = 0.48\,V.$$

3.5.2

a. Drain and gate of the transistor Q2 are shortly connected therefore it is in the saturation mode, so, $V_{inB} = V_{outB} + V_{T1}$. As the currents through the transistors are equal,

$$i_{D1} = K_n' \left(\frac{W}{L}\right)_1 (V_{G1B} - V_{T1})$$

$$= K_n' \left(\frac{W}{L}\right)_2 (V_{dd} - V_{outB} - V_{T2}) = i_{D2}$$

$$K_n' \left(\frac{W}{L}\right)_1 (V_{inB} - 1) = K_n' \left(\frac{W}{L}\right)_2 (10 - (V_{inB} - 1) - 1)$$

$$\left(\frac{W}{L}\right)_1 (V_{inB} - 1) = \left(\frac{W}{L}\right)_2 (10 - V_{inB});$$

$$V_{inB} = 1.9\,V, V_{outB} = 0.9\,V.$$

b. Calculation of the surface area.
$S_1 = W_1 \cdot L_1 = 100 \cdot 6 = 600\,\mu m^2$
$S_2 = W_2 \cdot L_2 = 6 \cdot 30 = 180\,\mu m^2$
$S_\Sigma = S_1 + S_2 = 600 + 180 = 780\,\mu m^2.$

3.5.3 Firstly, we will check the resistances of the transistors, the driver and the load.

a.
$$R_D = \frac{1}{K_n' \left(\frac{W}{L}\right)_n (V_{DD} - V_{TN})}$$

$$= \frac{1}{50 \cdot 10^{-6}(5 - 0.8)4/2} = 2.4\,k\Omega$$

$$R_L = \frac{1}{K_n' \left(\frac{W}{L}\right)_p (V_{DD} - V_{TN})}$$

$$= \frac{1}{50 \cdot 10^{-6}(5 - 0.8)10/2} = 2.9\,k\Omega.$$

As $R_D \neq R_L$, this circuit is not a CMOS inverter. The maximum output voltage is V_{DD} and the minimum output voltage is zero for this circuit. Now we can find V_{IL} and V_{IH}. At the point M when

$V_{\text{inM}} = V_{\text{IL}}$, the same current flows through both transistors: Q_D is in the saturation state and Q_L is in the triode state:

$$Kn(V_{\text{DSN}} - V_{\text{TN}})^2 = Kp[2(V_{\text{GSP}} - |V_{\text{TP}}|)V_{\text{DSP}} - V_{\text{DSP}^2}]$$

$$Kn(Vin_M - V_{\text{TN}})^2 = Kp[2(V_{\text{DD}} - Vin_M - |V_{\text{TP}}|)$$

$$\times (V_{\text{DD}} - V_{\text{outM}}) - (V_{\text{DD}} - V_{\text{outM}})^2]$$

$$2Kn(V_{\text{IL}} - V_{\text{TN}}) = Kp[-2(V_{\text{DD}} - V_{\text{outM}})$$

$$-2(V_{\text{DD}} - V_{\text{IL}} - |V_{\text{TP}}|)\frac{dV_{out}}{dV_{in}}$$

$$-2(V_{DD} - V_{out})\left(-\frac{dV_{out}}{dV_{in}}\right); \quad \frac{dV_{out}}{dV_{in}} = -1$$

$$Kn(V_{\text{IL}} - V_{\text{TN}}) = -Kp[(V_{\text{DD}} - V_{\text{outM}})$$

$$-2(V_{\text{DD}} - V_{\text{IL}} - |V_{\text{TP}}|) + (V_{DD} - V_{\text{outM}})$$

$$V_{outM} = \frac{1}{2}\left[\left(1 + \frac{K_n}{K_p}\right)V_{IL} + V_{DD} - \frac{K_n}{K_p}V_{TN} + |V_{TP}|\right]$$

$$V_{IL} = V_{TN} + \frac{V_{DD} - V_{TN} - |V_{TP}|}{\frac{K_n}{K_p} - 1}\left[2\sqrt{\frac{\frac{K_n}{K_p}}{\frac{K_n}{K_p} + 3}} - 1\right]$$

$$V_{IL} = 0.8 + \frac{5 - 0.8 - 0.8}{\frac{3 \cdot 4}{10} - 1}\left[2\sqrt{\frac{1.2}{1.2 + 3}} - 1\right] = 1.97\,V$$

$$V_{IH} = V_{TN} + \frac{V_{DD} - V_{TN} - |V_{TP}|}{\frac{K_n}{K_p} - 1}\left[\frac{2\frac{K_n}{K_p}}{\sqrt{3\frac{K_n}{K_p} + 1}} - 1\right]$$

$$= 0.8 + \frac{5 - 0.8 - 0.8}{1.2 - 1}\left[2\sqrt{\frac{1.2}{1.2 + 3}} - 1\right] = 2.82\,V$$

$$N_{MH} = V_{OH} - V_{IH} = 5 - 2.82 = 2.18\,V$$

$$N_{ML} = V_{IL} - V_{OL} = 1.97\,V.$$

b. An input voltage equal to 0 or V_{DD} does not produce a current in the circuit as, in both these states, one of the transistors is in the cutoff mode. If Vin $= V_{DD}/2$,

$$i = \mu_p C_{ox} \left(\frac{W}{L}\right)_p (V_{DD} - V_{in} - |V_{TP}|)^2$$

$$= \frac{50 \cdot 10^{-6}}{3} \frac{10}{2} (5 - 2.5 - 0.8)^2 = 0.24 \, mA.$$

c. The propagation delay t_{PHL} is the time it takes to transit from the high state to the low state $t_{PHL} = \frac{C\Delta V}{i_{av}}$, here i_{av} is the average current through the transistor Q_D in two different modes: triode and saturation. In other words, this current is the discharge current of the capacitor C.

$$i_{av} = \frac{i_{N0} + i_{N1}}{2}, \quad \Delta V = \frac{V_{DD}}{2}$$

$$i_{N0} = \frac{1}{2} K'_n \left(\frac{W}{L}\right)_n (V_{DD} - V_{TN})^2$$

$$= 0.5 \cdot 50 \cdot 10^{-6} \cdot 2 \cdot (5 - 0.8)^2 = 0.88 \, mA$$

$$i_{N1} = K'_n \left(\frac{W}{L}\right)_n \left[(V_{DD} - V_{TN})\frac{V_{DD}}{2} + \frac{1}{2}\left(\frac{V_{DD}}{2}\right)^2\right] = 1.36 \, mA$$

$$i_{av} = 0.5(0.88 + 1.36) = 1.12 \, mA,$$

$$t_{PHL} = \frac{0.1 \cdot 10^{-12} 2.5}{1.12 \cdot 10^{-3}} = 0.22 \, s.$$

3.5.4 The output equation is linear (see Eq. (3.24)).
a. $V_{out} = V_{DD} - i_{RD} R_D = V_{DD} - (i_{AD} + i_{BD})R_D = V_{DD} - 2R_D \cdot 0.5K[2(V_{DD} - V_T)V_{out} - V_{out}^2]$.
Here, V_{out} is small, therefore V_{out}^2 may be neglected.

$$V_{out} \approx = V_{DD} - 2R_D \cdot K(V_{DD} - V_T)V_{out}$$

$$V_{out} = \frac{V_{DD}}{1 + 2KR_D(V_{DD} - V_T)}$$

$$= \frac{5}{1 + 2 \cdot 2 \cdot 50 \cdot 10^{-6} \cdot 50 \cdot 10^3 \cdot 4} 0.12 \, V.$$

Capacitor C charges through the resistance R_D up to the voltage $0.9V_{DD}$:

$V = V_{DD}(1 - e^{\frac{t}{\tau}})$ where $\tau = RC$, so $t = -RCln(1 - 0.9) = 2.3\,RC$.

The charge time $t_{PLH} = 2.3\,R_D C = 2.3 \cdot 50 \cdot 10^3 \cdot 5 \cdot 10^{-12} = 0.58\,\mu s$.

The capacitor discharges through one or two transistors, therefore the resistance of one transistor will be (see Eq. (3.23)) as follows:

$$R_{on} = \frac{1}{K'_n \left(\frac{W}{L}\right)_n (V_{DD} - V_T)} = \frac{1}{2 \cdot 50 \cdot 10^{-6}\,(5 - 1)} = 2.5\,k\Omega$$

$t_{PHL} = 2.3 R_{on} C = 2.3 \cdot 2.5 \cdot 10^3 \cdot 5 \cdot 10^{-12} = 28.8\,ns.$

3.5.5 Here both transistors are in the saturation state due to their diode-type connection. Therefore, the current through the circuit is equal to:

$$i = \frac{1}{2}(\mu_n C_{ox}) \left(\frac{W}{L}\right)_n (V_{GS} - V_T)^2$$

$Q1 : 200 = 0.5 \cdot 20 \cdot 0.1 \cdot W_1 \cdot (3 - 2) \Longrightarrow W_1 = 200\,\mu m$

$Q2 : 200 = 0.5 \cdot 20 \cdot 0.1 \cdot W_2 \cdot (4 - 2) \Longrightarrow W_2 = 50\,\mu m$

$R = (10 - 7)/0.2 \cdot 10^{-3} = 15\,k\Omega.$

3.5.6 This circuit represents the SR latch. Assume that the initial state of the latch is:

$$Q = \text{``1''} \quad \text{and} \quad \bar{Q} = \text{``0''}.$$

To transit the circuit into the complement state, we need to apply the following signals to the inputs: $S = $ "0" and $R = $ "1". In this case, the transistor T5 will be open and the current through transistors T3-T5 begins to flow. In the initial stage T5 will be in the saturation mode and T3 will be in the triode mode.

$$K_5(V_R - V_T)^2 = K_3\left[2(V_{\bar{Q}} - V_T)V_Q - \frac{1}{2}V_{\bar{Q}}^2\right]$$

$$V_R = V_Q = \frac{V_{DD}}{2}; \; K_5(2.5 - 1)^2 = K_3[2(5 - 1)2.5 - 2.5^2]$$

$$K_5 = K_6 = K\frac{2.4 \cdot 2.5 - 2.5^2}{1.5^2} = 6.11K.$$

3.5.7 To identify the logic function, we use the analytical method.

a. $V_{out} = \overline{AB + V1} = \overline{AB + \overline{AC + B}} = \overline{AB + \bar{B}(\bar{A} + \bar{C})}$

$$= \overline{AB + \bar{A}\bar{B} + \bar{B}\bar{C}} = (\bar{A} + \bar{B})(A + B)(B + C) = \bar{A}B + A\bar{B}C.$$

A truth table (Table 6.7) built for the circuit will help us to find out what the input's states for minimum V_{out} are.

b. For calculation according to Fig. 6.4, we can choose the state when only transistor $Q3$ will be conductive. In this case, both the inputs A and B should be in the low state.

Table 6.7.   Truth table.

A	B	C	V_{out}
0	0	0	0
0	0	1	0
0	1	0	1
0	1	1	1
1	0	0	0
1	0	1	1
1	1	0	0
1	1	1	0

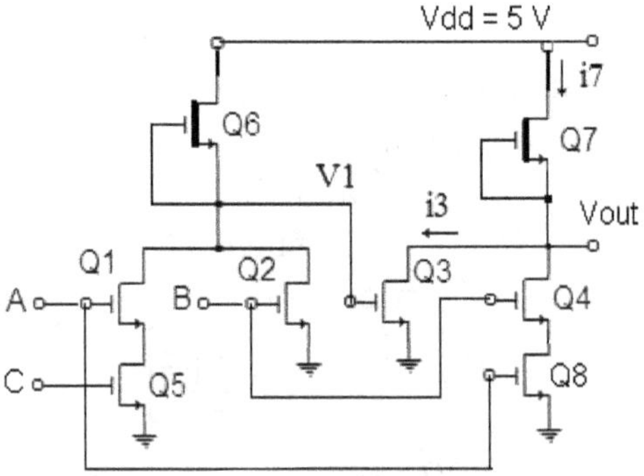

Fig. 6.4.   Logic gate.

Now, as shown in the figure, we have two equal currents $i3 = i7$ when $Q3$ is in the triode mode and $Q7$ is in the saturation mode.

$$\frac{1}{2}K'_n\left(\frac{W}{L}\right)_7 (V_{GS7} - |V_{TN}|)^2$$

$$= K'_n\left(\frac{W}{L}\right)_3 \left[(V_{GS3} - V_{TN})V_{DS3} - \frac{1}{2}\left(\frac{V_{DS3}}{2}\right)^2\right]$$

$$\left(\frac{W}{L}\right)_7 (-|V_{TN7}|)^2 = \left(\frac{W}{L}\right)_3 [2(V_{GS3} - V_{TN})V_{oL} - V_{oL}^2]$$

$$2V_{oL}^2 - 16V_{oL} + 3 = 0; \quad V_{oL} = 0.19\,V.$$

3.5.8 Figure 6.5 represents the studied inverter, its V-A characteristic and the transfer function diagram.

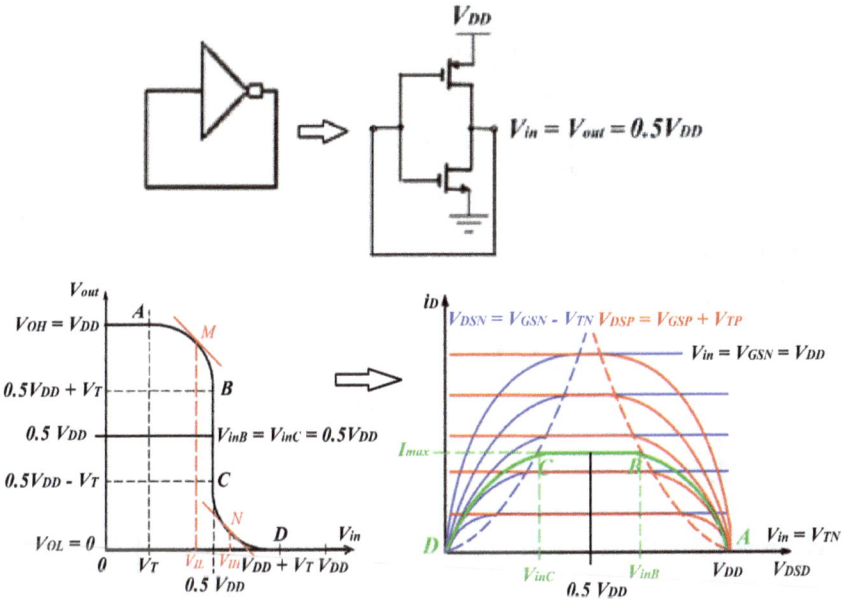

Fig. 6.5. The CMOS inverter with connected input and output and its characteristics.

Here, the input voltage is equal to the output voltage, therefore the circuit is in the middle of the transition process. The current which flows at this point is at its maximum:

$$I_{max} = \frac{1}{2}\left(\mu_n C_{ox}\right)\left(\frac{W}{L}\right)_n (0.5V_{DD} - V_T)^2$$

$$= 0.5 \cdot 25 \cdot 10^{-6} \cdot 1.6 \cdot (2.5 - 1)^2 = 45\,\mu A.$$

4.5.1 To identify the logic function, we use the analytical method. Figure 6.6 represents the logic circuit with suitable logic functions.

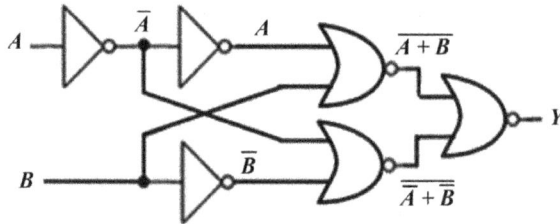

Fig. 6.6.   Digital logic circuit.

a. $Y = \overline{(A+B) + (\overline{A}+\overline{B})} = \overline{(\overline{AB}) + AB} = (A+B)(\overline{A}+\overline{B}) = A\overline{B} + \overline{A}B = A \oplus B.$

b. Figure 6.7 represents the CMOS circuit implementing the function defined in the previous part a.

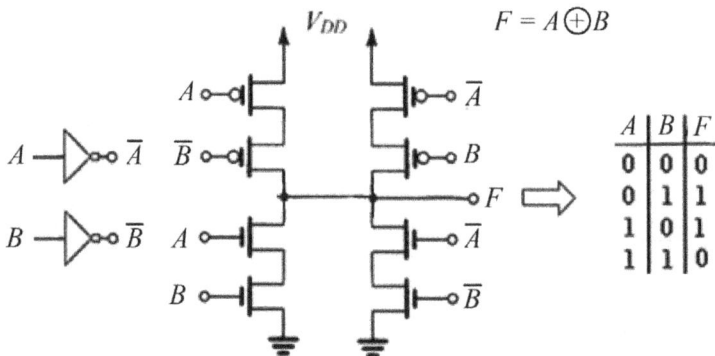

$F = A \oplus B$

A	B	F
0	0	0
0	1	1
1	0	1
1	1	0

Fig. 6.7.   A CMOS circuit implementing the XOR function.

c. Figure 6.8 represents the digital circuit implementing the same function using PTL-CMOS technology.

$$F = A\bar{B} + \bar{A}B$$

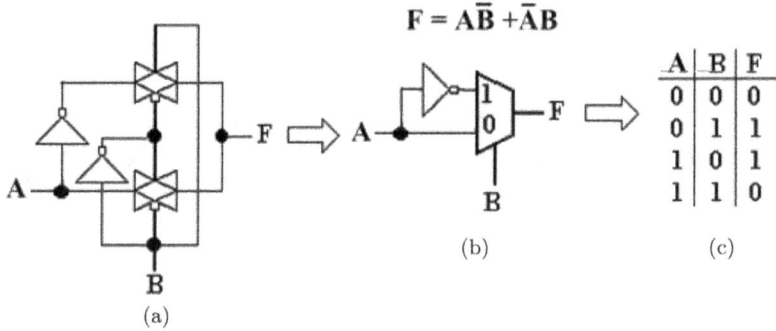

A	B	F
0	0	0
0	1	1
1	0	1
1	1	0

(a)   (b)   (c)

Fig. 6.8.   A PTL-CMOS implementation of the XOR function.

d. $S_{CMOSe} = S_{Ne} + S_{Pe}; \; S_{Ne} = LW_N = L^2 \left(\dfrac{W}{L} \right)_N$

$$= 2^2 \cdot 1.5 = 6 \, \mu m^2$$

$$S_{Pe} = L \cdot W_P = L \cdot W_N \left(\frac{\mu_N}{\mu_P} \right) = L^2 \left(\frac{W}{L} \right)_N \left(\frac{\mu_N}{\mu_P} \right)$$

$$= 2^2 \cdot 1.5 \cdot 2.5 = 15 \, \mu m^2$$

$$S_{CMOSe} = 21 \, \mu m^2$$

$$S_{PTL-CMOS} = 4 \cdot S_{CMOSe} = 4 \cdot 21 = 84 \, \mu m^2$$

$$S_{CMOS} = 2 \cdot S_{CMOSe} + 4 \cdot 2 \cdot S_{Pe} + 4 \cdot 2 \cdot S_{Ne} = 210 \, \mu m^2.$$

Table 6.8.   Truth table.

A	B	C	F
0	0	0	1
0	0	1	1
0	1	0	0
0	1	1	1
1	0	0	1
1	0	1	1
1	1	0	1
1	1	1	1

4.5.2 To identify the logic function, we use the analytical method and the truth table (see Table 6.8).

a. $\mathbf{F} = \{[(AB)'' + C]'(B + C)\}' = [(AB + C)'(B + C)]' = [(A' + B')C'(B + C)]' = [(A' + B')C'B + (A' + B')C'C]' = [(A'BC')]' = A + B' + C; \mathbf{F}' = \mathbf{A'BC'}.$

Figure 6.9 represents the CMOS (Fig. 6.9(a)) and PTL-CMOS (Fig. 6.9(b)) implementations of the studied logic function.

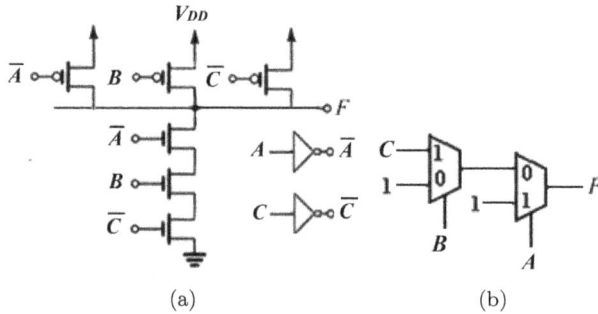

Fig. 6.9.   CMOS implementation (a) and PTL-CMOS implementation (b) of the defined logic function.

b.
$$S_{CMOSe} = S_{Ne} + S_{Pe}; \; S_{Ne} = LW_N = L^2 \left(\frac{W}{L}\right)_N$$

$$= 2^2 \cdot 1.5 = 6 \, \mu m^2$$

$$S_{Pe} = L \cdot W_P = L \cdot W_N \left(\frac{\mu_N}{\mu_P}\right) = L^2 \left(\frac{W}{L}\right)_N \left(\frac{\mu_N}{\mu_P}\right)$$

$$= 2^2 \cdot 1.5 \cdot 2.5 = 15 \, \mu m^2$$

$$S_{CMOSe} = 21 \, \mu m^2$$

$$S_{PTL-CMOS} = 6 \cdot S_{CMOSe} = 6 \cdot 21 = 126 \, \mu m^2$$

$$S_{CMOS} = 2 \cdot S_{CMOSe} + 3 \cdot S_{Pe} + 3 \cdot 3 \cdot S_{Ne} = 141 \, \mu m^2.$$

c. The circuit designed in part b is the CMOS circuit, therefore the propagation delay is:

$$t_{PHL} = t_{PLH} = \frac{1.6C}{K'_n \left(\frac{W}{L}\right)_n V_{DD}} = \frac{1.6 \cdot 10 \cdot 10^{-12}}{5 \cdot 10^{-6} 1.5 \cdot 5} = 0.43 \, \mu s.$$

4.5.3 To identify the logic function, we use the truth table (Table 6.9).

Table 6.9. Truth table.

F(t)	A	C	F(t+1)
0	0	0	0
0	0	1	0
0	1	0	0
0	1	1	1
1	0	0	1
1	0	1	0
1	1	0	1
1	1	1	1

a.

	00	01	11	10
0	0	0	1	0
1	1	0	1	1

The Karnaugh map helps us to define the required function.

$$F(t+1) = AC + F(t)C' = [(A' + C')(F'(t) + C)]'.$$

b. Figure 6.10 represents the built electrical circuits.

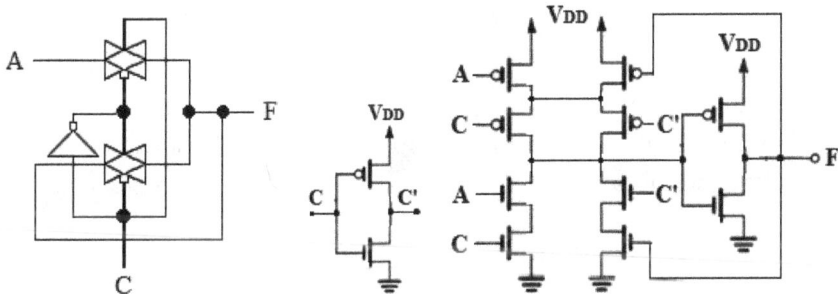

Fig. 6.10. The PTL-CMOS and CMOS implementations of the logic function.

4.5.4 To identify the logic function, we use the truth table method.

a. $D = (A+B) \otimes Y'_{(t)}$; $Y_{(t+1)} = D \cdot Clk$

D	Y(t+1)
0	0
1	1

b. Figure 6.11 represents the PTL-CMOS implementation of the synchronous D flip-flop.

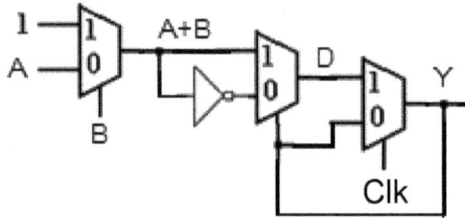

Fig. 6.11.   The PTL-CMOS implementation of the synchronous D flip-flop.

c. Figure 6.12 represents the CMOS implementation of the synchronous D flip-flop.

4.5.5 It is known that the propagation delay of the CMOS inverter is (see Eq. (4.38)):

$$T = 2nt_{PHL}, \quad t_s PHL = \frac{1}{2nf} = \frac{1}{2 \cdot 5 \cdot 10^8} = 10^{-7}s.$$

According to the assumption that $V_T = 0.2V_{DD}$, and Eq. (3.92) for the maximum current flowing through the CMOS inverter under transition, we obtain:

$$i_D = 0.5K_n(0.5\,V_{DD} - V_T)^2,$$

$$K_n = \frac{2i_D}{(0.3V_{DD})^2} = \frac{2 \cdot 0.2}{(0.3 \cdot 1.8)^2} = 1.37\,A/V^2.$$

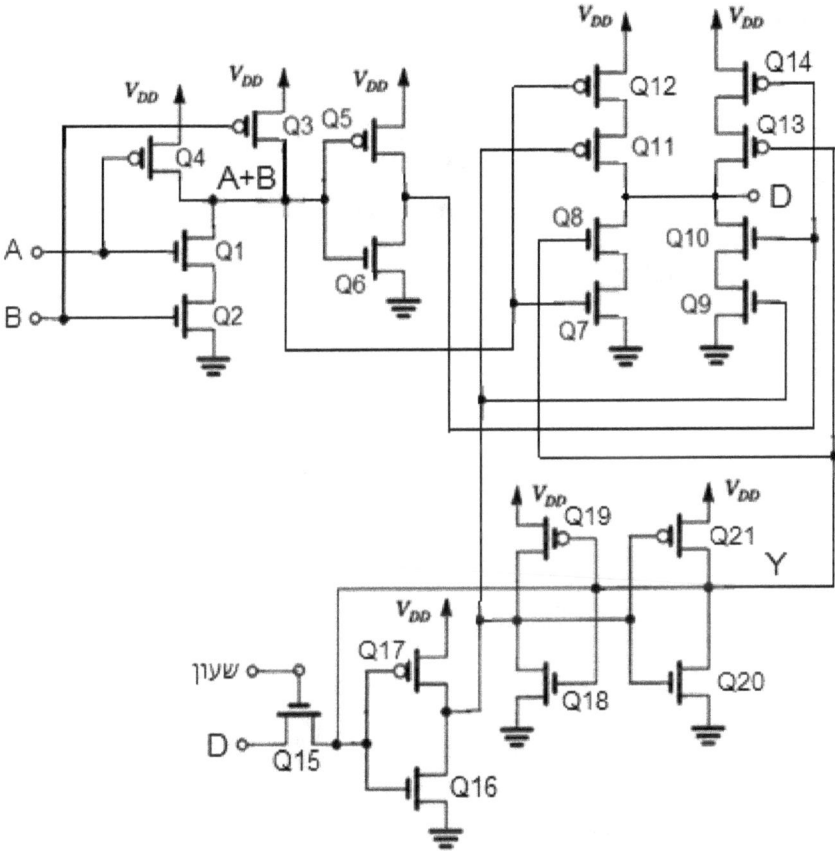

Fig. 6.12. The CMOS implementation of the synchronous D flip-flop.

The input capacity of the CMOS inverter may be found from Eq. (3.90):

$$t_{PHL} = \frac{1.6C}{K_n V_{DD}}, \quad C = \frac{t_{PHL} K_n V_{DD}}{1.6} = \frac{10^{-7} 1.37 \cdot 1.8}{1.6} = 0.15\,\mu F.$$

The power of each inverter is:

$$P = \frac{1}{n} f C V_{DD}^2 = 0.2 \cdot 10^8 \cdot 0.15 \cdot 10^{-6} 1.8^2 = 9.72\,W.$$

4.5.6 Figure 6.13 illustrates the behavior of the multivibrator.

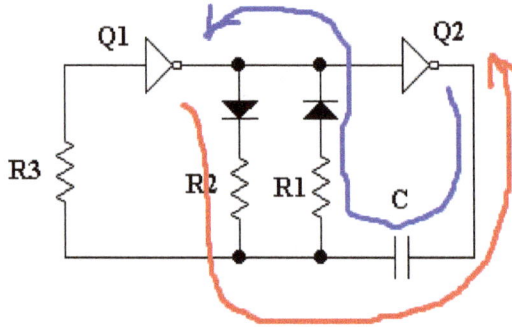

Fig. 6.13.   Multivibrator behavior.

The output voltages of a CMOS inverter in the stable state are V_{DD} and zero. If the DC = 0.2, so T1 + T2 = T = 50 ms, T1 = 0.2T = 10 ms and T2 = 40 ms.

$$i_c = I_0 e^{-\frac{t}{\tau}} = \frac{V_{DD} - V_{D,on}}{R1} e^{-\frac{T1}{R1C}}$$

$$V_{Th} = \frac{V_{DD}}{2} = V_{DD} - V_{D,on} - i_c R1$$

$$(V_{DD} - V_{D,on}) e^{-\frac{T1}{R1C}} = \frac{V_{DD}}{2} - V_{D,on}$$

$$T1 = -R1Cln\left(\frac{0.5V_{DD} - V_{D,on}}{V_{DD} - V_{D,on}}\right)$$

$$= -R1 \cdot 10^{-6} ln\frac{1.8}{4.3} = 10 \cdot 10^{-6}, \ R1 = 11.5\,\Omega$$

$$T2 = 40 \cdot 10^{-6} = -R2 \cdot 10^{-6} ln\frac{1.8}{4.3}, \ R2 = 46\,\Omega.$$

4.5.7

a. Figure 6.14 implements the shift register function.

b.

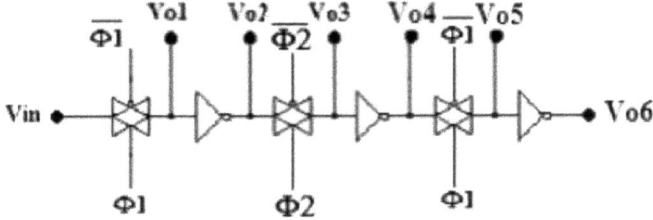

Fig. 6.14.   Shift register circuit.

c. Propagation time calculation. One stage of the register is shown
in the following figure, Fig. 6.15.

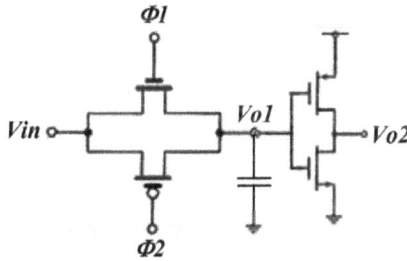

Fig. 6.15.   One stage of the shift register.

If we assume that $|V_T| = 1\,\text{V}$ and $V_{DD} = 5\,\text{V}$, the propagation
time will be a sum of transitions through the switch and inverter:
$T = T1 + T2$.

$$T1 = \frac{CV_{DD}}{2i_{Dp}}, \quad T2 = \frac{1.6C}{K_n'\left(\frac{W}{L}\right)_n V_{DD}},$$

$$i_{Dp} = \frac{1}{2}K_p(V_{DD} - |V_T|)^2, \quad K_n = K_p = 30\frac{\mu A}{V^2}$$

274	Digital Electronic Circuits: The Comprehensive View

$$W_p = \frac{\mu_n}{\mu_p} W_n = \frac{\mu_n}{\mu_p} \left(\frac{W}{L}\right)_n L = 3 \cdot 2 \cdot 2 = 12 \, \mu m$$

$$T1 = \frac{C V_{DD}}{2\frac{1}{2} K_p' \left(\frac{W}{L}\right)_p (V_{DD} - |V_T|)^2} = \frac{10 \cdot 10^{-15} 5}{30 \cdot 10^{-6} \cdot 6 \cdot 4^2} = 0.02 \, ns$$

$$T2 = \frac{1.6 \cdot 10 \cdot 10^{-15}}{30 \cdot 2 \cdot 5 \cdot 10^{-6}} = 0.05 \, ns,$$

$$T = 0.02 + 0.05 = 0.07 \, ns, \quad T3 = T \cdot 3 = 0.21 \, ns.$$

4.5.8

a. The circuit's behavior is illustrated in Table 6.10.

Table 6.10. Logic circuit behavior.

A	B	LED
Low	Low	Off
Low	High	On
High	Low	Off

b. If A = D = High, we obtain the oscillating simplified circuit shown in Fig. 6.16.

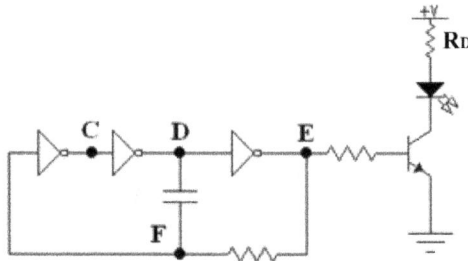

Fig. 6.16.   Oscillating simplified circuit.

c. $R_D = (V_{DD} - V_{LED} - V_{CE,sat})/I_D = (12 - 1.5 - 0.2)/15 \approx 690\Omega.$
d. The dynamic behavior of the circuit is illustrated by the timing diagram shown in Fig. 6.17.

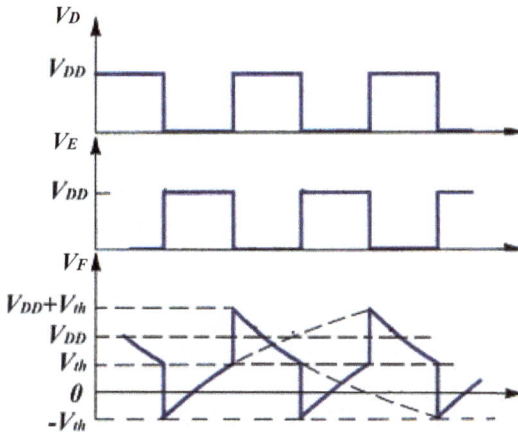

Fig. 6.17.   Timing diagram.

4.5.9 This is a one-shot circuit.

a. In this circuit, transistors Q5 and Q6 play the role of resistors:

$$R_{PMOS} = \frac{1}{k_p' \left(\frac{W}{L}\right)_p (V_{dd} - |V_T|)};$$

$$R_5 = \frac{1}{40 \cdot 10^{-6} \cdot \left(\frac{1.2}{8}\right) \cdot (3 - 0.6)} = 69.4\,k\Omega$$

$$R_1 = \frac{1}{100 \cdot 10^{-6} \cdot \left(\frac{1.2}{0.8}\right) \cdot (3 - 0.6)} = 28\,k\Omega$$

$$T = \mathbf{C}\,(R_1 + R_2)\,ln\left(\frac{R_5}{R_1 + R_5} \cdot \frac{V_{dd}}{V_{dd} - V_{Th}}\right)$$

$$T = \mathbf{10} \cdot 10^{-12}\,(69.4 + 2.8)\cdot 10^3 ln\left(\frac{69.4}{69.4 + 2.8} \cdot \frac{3}{3 - 1.5}\right)$$

$$= 0.48\,\mu s.$$

b. The width of the input pulse should be more than the transition time of the inverter $Q3 - Q4$. As these transistors are complementary, one can use the propagation time equation for CMOS:

$$T_{in,min} \geq t_{PHL} = \frac{1.6C}{k_n' \left(\frac{W}{L}\right)_n V_{dd}} = \frac{1.6 \cdot 10 \cdot 10^{-12}}{100 \cdot 10^{-6} \cdot 1.5 \cdot 3} = 36\,ns.$$

c. The maximum current flows through the circuit $V_{DD} \rightarrow Q5 \rightarrow Q1 \rightarrow \perp$:

$$I_{max} = \frac{V_{dd}}{R_1 + R_5} = \frac{3}{(69.4 + 2.8) \cdot 10^3} = 416\,\mu A.$$

4.5.10

a. The logic function of the multiplexer is shown in the truth table (Table 6.11).

Table 6.11. Truth table of the multiplexer.

X	Y	F
0	0	A
0	1	B
1	0	C
1	1	D

$$F = \bar{X}\bar{Y}A + \bar{X}YB + X\bar{Y}C + XYD$$
$$\bar{F} = (X + Y + \bar{A})(X + \bar{Y} + \bar{B})(\bar{X} + Y + \bar{C})(\bar{X} + \bar{Y} + \bar{D}).$$

b. The PTL-CMOS circuit implementation is shown in Fig. 6.18.

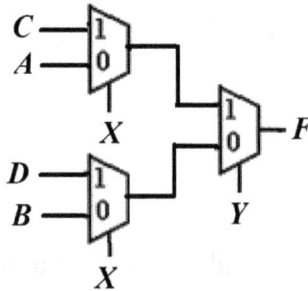

Fig. 6.18. The PTL-CMOS implementation of the multiplexer.

c. The CMOS circuit is shown in Fig. 6.19.

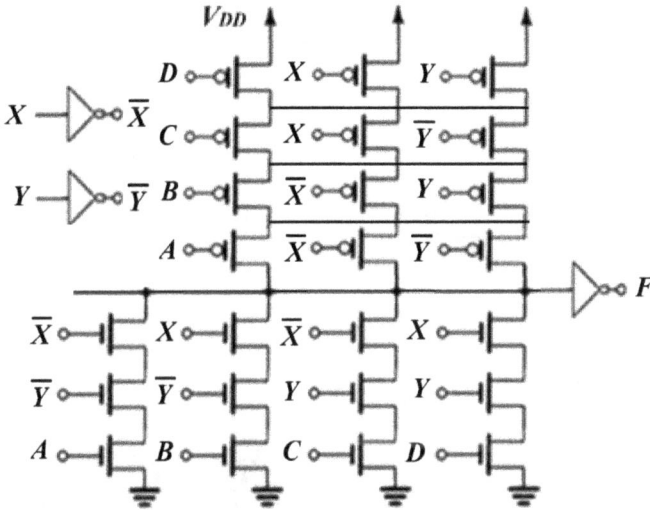

Fig. 6.19.   The CMOS implementation of the multiplexer.

d. The calculation of the circuit areas.

$$S_{CMOSe} = S_{Ne} + S_{Pe}; \quad S_{Ne} = LW_N = L^2 \left(\frac{W}{L}\right)_N$$

$$= 2^2 \cdot 2 = 8\,\mu m^2$$

$$S_{Pe} = L \cdot W_P = L \cdot W_N \left(\frac{\mu_N}{\mu_P}\right) = L^2 \left(\frac{W}{L}\right)_N \left(\frac{\mu_N}{\mu_P}\right)$$

$$= 2^2 \cdot 2 \cdot 2.5 = 20\,\mu m^2$$

$$S_{CMOSe} = 28\,\mu m^2$$

$$S_{PTL-CMOS} = 9 \cdot S_{CMOSe} = 9 \cdot 28 = 252\,\mu m^2$$

$$S_{CMOS} = 3 \cdot S_{CMOSe} + 12 \cdot 4 \cdot S_{Pe} + 12 \cdot 3 \cdot S_{Ne} = 1332\,\mu m^2.$$

5.6.1 The leakage current may be calculated from the general equation for the charge:

$$Q = CV = IT, \quad I = \frac{CV}{T} = \frac{0.1 \cdot 10^{15} \cdot 1.2}{10^{-3}} = 0.12\,nA.$$

5.6.2 To program the MOS ROM, we need to build a truth table (Table 6.12).

Table 6.12.   Truth table.

A	8	0	2	B	F	4	5	
0	0	0	0	1	1	0	1	**X0**
1	0	0	1	1	1	0	0	**X1**
0	0	0	0	0	1	1	1	**X2**
1	1	0	0	1	1	0	0	**X3**
7	6	5	4	3	2	1	0	

Figure 6.20 represents the MOS-ROM electrical circuit built according to Table 6.12.

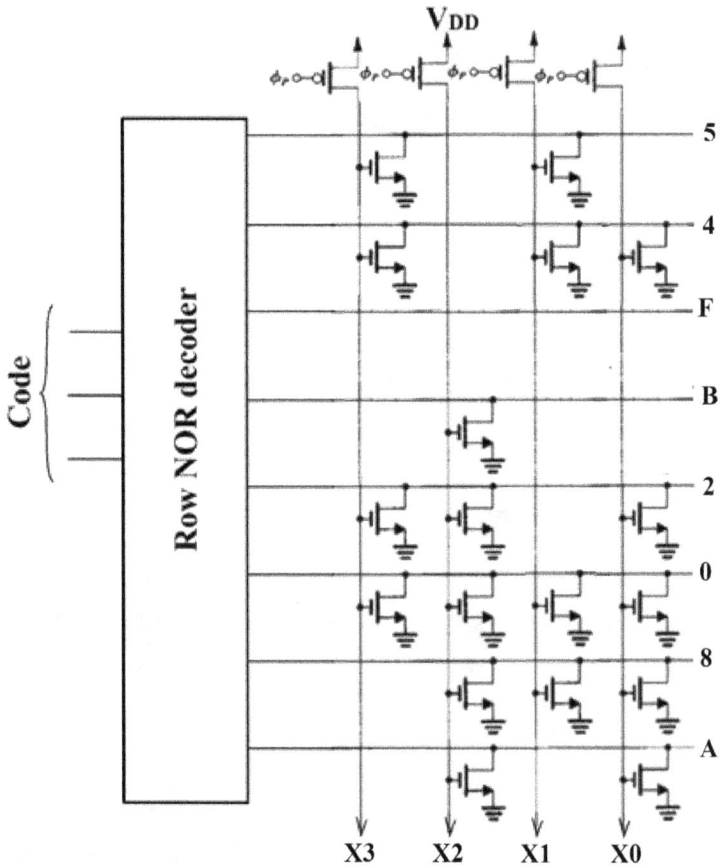

Fig. 6.20.   The electrical circuit of the MOS-ROM.

5.6.3 In this circuit, Fig. 6.21, the bit-lines are precharged up to high voltage, the cell contains low voltage and the complementary contact $\bar{Q} = 1 = V_{DD}$. Therefore, after applying the voltage to the word-line that is the voltage on the gates of transistors $Q5$ and $Q6$, the current will flow only in the circuit shown by the red color (VB→ $Q6 \to Q3$ → ⊥) as presented in the following picture. Here, transistor $Q5$ is in the saturation state and transistor $Q3$ is in the triode state and they have a common current $i_6 = i_3$.

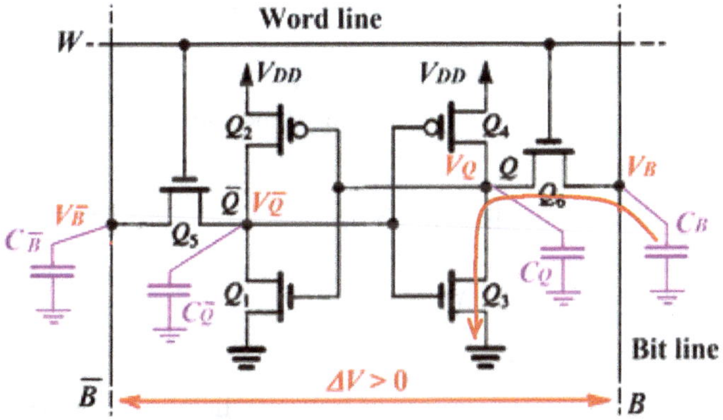

Fig. 6.21. A SRAM memory cell.

$$i_3 = (\mu_n C_{ox}) \left(\frac{W}{L}\right)_n \left[(V_{DD} - V_{TN0}) V_Q - \frac{1}{2} V_Q^2\right]$$

$$= \frac{1}{2} (\mu_n C_{ox}) \left(\frac{W}{L}\right)_{PTL,n} (V_{DD} - V_Q - V_T)^2 = i_6$$

$$40 \cdot 10^{-6} \cdot 2 \left[(5-1) V_Q - \frac{1}{2} V_Q^2\right] = \frac{1}{2} 40 \cdot 10^{-6} (5 - V_Q - 1)^2$$

$$3V_Q^2 - 24V_Q + 16 = 0, \quad V_Q = 0.73 \, V.$$

Evidently, only one of the two solutions is right, as the second is more than V_{DD}.

5.6.4 Memory information is stored in the virtual capacitor C_Q, as shown in Fig. 6.22.

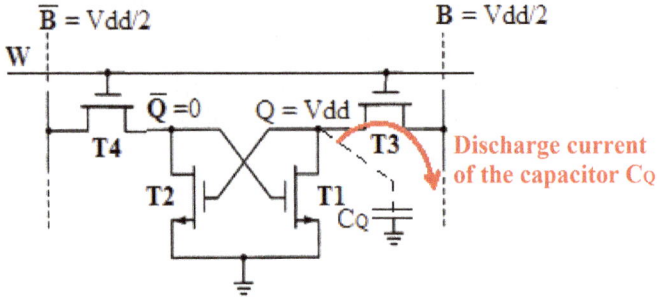

Fig. 6.22. The dynamical memory cell.

The capacitance of this capacitor:

$$C_Q = C_{ox}WL = \frac{k'_n}{\mu_n}\left(\frac{W}{L}\right)_n L^2 = \frac{40 \cdot 10^{-6} \cdot 2 \cdot 0.6^2 \cdot 10^{-8}}{1000} \approx 0.3\, fF.$$

When the row is selected, currents begin to flow in both parts of this symmetrical circuit.

$$I_3 = 0.5k'_n\left(\frac{W}{L}\right)_n (V_{dd} - V_{TN})^2$$

$$= 0.5 \cdot 40 \cdot 10^{-6} \cdot 2 \cdot (4 - 0.8)^2 = 0.41\, mA.$$

As this current is constant, we obtain:

$$t = \frac{C_Q \cdot \Delta V}{I_3} = \frac{0.3 \cdot 10^{-15} \cdot 0.25 \cdot 10^{-3}}{0.41 \cdot 10^{-3}} = 0.18\, fs.$$

5.6.5 Memory information is stored in the virtual capacitor C_Q, as shown in Fig. 6.23.

$$C_Q = C_{ox}WL = \frac{k'_n}{\mu_n}\left(\frac{W}{L}\right)_n L^2 = \frac{20 \cdot 10^{-6} \cdot 2.5 \cdot 0.6^2 \cdot 10^{-8}}{1000} = 0.18 fF.$$

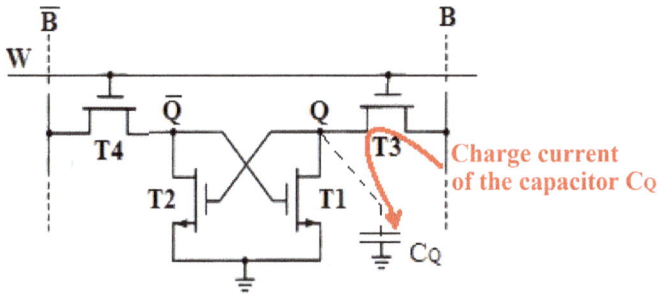

Fig. 6.23. The dynamic memory cell.

When the row is selected, a current begins to flow through the transistor T3 and charges the virtual capacitor C_Q. Transistor T3 is in the saturation state at the initial time of the charging process and it ends this process in the triode mode.

$$i_D = \frac{i_1 + i_2}{2} = \frac{\frac{1}{2}K_n \left(V_{DD} - V_T\right)^2 + \frac{1}{2}\left[2\left(V_{DD} - V_T\right)V_Q - V_Q^2\right]}{2}$$

$$= K'_n \left(\frac{W}{L}\right)_n \frac{\left(V_{DD} - V_T\right)^2 + \left[2\left(V_{DD} - V_T\right)V_T - V_T^2\right]}{4}$$

$$20 \cdot 10^{-6} \cdot 2.5 \frac{\left(2 - 0.4\right)^2 + \left[2\left(2 - 0.4\right)0.4 - 0.4^2\right]}{4} = 33.5 \cdot 10^{-6} A.$$

Charging through the NMOS transistor can reach only the decreased voltage $(V_{DD} - V_T)$.

$$\Delta t = \frac{C_Q \Delta V}{i_D} = \frac{0.18 \cdot 10^{-15} \left(2 - 0.4\right)}{33.5 \cdot 10^{-6}} = 8.6 \cdot 10^{-12} s.$$

Bibliography

Beards, Peter H., *Analog and Digital Electronics, A First Course*, Prentice Hall, New York, 1991.

Cahill, S. J., *Digital and Microprocessor Engineering*, Ellis Horwood, New York, 1993.

Hamacher C., Z. Vranesic, and S. Zaky, *Computer Organization*, McGraw Hill, New York, 2002.

Hennessy J. and D. Patterson, *Computer Architecture, A Quantitative Approach*, Kaufmann, Amsterdam, 1996.

Horowitz P. and W. Hill, *The Art of Electronics*, Cambridge University Press, Cambridge, 1989.

Fjeldly, Tor A., T. Ytterdal and M. S. Shur, *Introduction to Device Modeling and Circuit Simulation*, Wiley, New York, 1998. See also: https://ecse.rpi.edu/~shur/Ch5/index.htm.

Gingrich, D. M., *Physics Lecture Notes*, University of Alberta, Department of Physics, 1999. See also: http://www.piclist.com/images/ca/ualberta/phys/www/http/~gingrich/phys395/notes/phys395.html.

Kasap, S.O., *Principles of Electrical Engineering Materials and Devices*, McGraw-Hill, Boston, 1997.

Mano, M.M., *Computer System Architecture*, Prentice Hall, New Jersey, 1993.

Mano, M.M., *Digital Design*, Prentice Hall, New Jersey, 2002.

Neamen, Donald A., *Microelectronics: Circuit Analysis and Design*, McGraw Hill, 3rd Ed., 2007.

Sedra, Adel S. and Kenneth C. Smith, *Microelectronic Circuits*, Oxford University Press, 6th Ed., 2010.

Shiva, Sajjan G., *Computer Organization, Design, and Architecture*, CRC-Press, 4th Ed., 2008.

Singh, Jasprit, *Semiconductor Devices: Basic Principles*, JohnWiley, 2001. See also http://www.eecs.umich.edu/courses/eecs320/.

Van Zeghbroeck, Bart J., *Principles of Semiconductor Devices*, University of Colorado at Baulder, 1999. See also: http://ece-www.colorado.edu/~bart/book/.

Wakerly J.F., *Digital Design, Principle and Practice*, Prentice Hall, New Jersey, 2001.

Index